WHAT EVERY ENGINEER SHOULD KNOW ABOUT SOFTWARE ENGINEERING

WHAT EVERY ENGINEER SHOULD KNOW
A Series

Series Editor*

Phillip A. Laplante

Pennsylvania State University

1. What Every Engineer Should Know About Patents, *William G. Konold, Bruce Tittel, Donald F. Frei, and David S. Stallard*
2. What Every Engineer Should Know About Product Liability, *James F. Thorpe and William H. Middendorf*
3. What Every Engineer Should Know About Microcomputers: Hardware/Software Design, A Step-by-Step Example, *William S. Bennett and Carl F. Evert, Jr.*
4. What Every Engineer Should Know About Economic Decision Analysis, *Dean S. Shupe*
5. What Every Engineer Should Know About Human Resources Management, *Desmond D. Martin and Richard L. Shell*
6. What Every Engineer Should Know About Manufacturing Cost Estimating, *Eric M. Malstrom*
7. What Every Engineer Should Know About Inventing, *William H. Middendorf*
8. What Every Engineer Should Know About Technology Transfer and Innovation, *Louis N. Mogavero and Robert S. Shane*
9. What Every Engineer Should Know About Project Management, *Arnold M. Ruskin and W. Eugene Estes*

*Founding Series Editor: **William H. Middendorf**

10. What Every Engineer Should Know About Computer-Aided Design and Computer-Aided Manufacturing: The CAD/CAM Revolution, *John K. Krouse*
11. What Every Engineer Should Know About Robots, *Maurice I. Zeldman*
12. What Every Engineer Should Know About Microcomputer Systems Design and Debugging, *Bill Wray and Bill Crawford*
13. What Every Engineer Should Know About Engineering Information Resources, *Margaret T. Schenk and James K. Webster*
14. What Every Engineer Should Know About Microcomputer Program Design, *Keith R. Wehmeyer*
15. What Every Engineer Should Know About Computer Modeling and Simulation, *Don M. Ingels*
16. What Every Engineer Should Know About Engineering Workstations, *Justin E. Harlow III*
17. What Every Engineer Should Know About Practical CAD/CAM Applications, *John Stark*
18. What Every Engineer Should Know About Threaded Fasteners: Materials and Design, *Alexander Blake*
19. What Every Engineer Should Know About Data Communications, *Carl Stephen Clifton*
20. What Every Engineer Should Know About Material and Component Failure, Failure Analysis, and Litigation, *Lawrence E. Murr*
21. What Every Engineer Should Know About Corrosion, *Philip Schweitzer*
22. What Every Engineer Should Know About Lasers, *D. C. Winburn*
23. What Every Engineer Should Know About Finite Element Analysis, *John R. Brauer*
24. What Every Engineer Should Know About Patents: Second Edition, *William G. Konold, Bruce Tittel, Donald F. Frei, and David S. Stallard*
25. What Every Engineer Should Know About Electronic Communications Systems, *L. R. McKay*
26. What Every Engineer Should Know About Quality Control, *Thomas Pyzdek*

27. What Every Engineer Should Know About Microcomputers: Hardware/Software Design, A Step-by-Step Example. Second Edition, Revised and Expanded, *William S. Bennett, Carl F. Evert, and Leslie C. Lander*
28. What Every Engineer Should Know About Ceramics, *Solomon Musikant*
29. What Every Engineer Should Know About Developing Plastics Products, *Bruce C. Wendle*
30. What Every Engineer Should Know About Reliability and Risk *Analysis, M. Modarres*
31. What Every Engineer Should Know About Finite Element Analysis: Second Edition, Revised and Expanded, *John R. Brauer*
32. What Every Engineer Should Know About Accounting and Finance, *Jae K. Shim and Norman Henteleff*
33. What Every Engineer Should Know About Project Management: Second Edition, Revised and Expanded, *Arnold M. Ruskin and W. Eugene Estes*
34. What Every Engineer Should Know About Concurrent Engineering, *Thomas A. Salomone*
35. What Every Engineer Should Know About Ethics, *Kenneth K. Humphreys*
36. What Every Engineer Should Know About Risk Engineering and Management, *John X. Wang and Marvin L. Roush*
37. What Every Engineer Should Know About Decision Making Under Uncertainty, *John X. Wang*
38. What Every Engineer Should Know About Computational Techniques of Finite Element Analysis, *Louis Komzsik*
39. What Every Engineer Should Know About Excel, *Jack P. Holman*
40. What Every Engineer Should Know About Software Engineering, *Phillip A. Laplante*

WHAT EVERY ENGINEER SHOULD KNOW ABOUT

SOFTWARE ENGINEERING

Phillip A. Laplante

CRC Press is an imprint of the
Taylor & Francis Group, an **informa** business

CRC Press
Taylor & Francis Group
6000 Broken Sound Parkway NW, Suite 300
Boca Raton, FL 33487-2742

Printed in the United States of America on acid-free paper
10 9 8 7 6 5 4 3 2 1

International Standard Book Number-10: 0-8493-7228-3 (Softcover)
International Standard Book Number-13: 978-0-8493-7228-5 (Softcover)

Library of Congress Cataloging-in-Publication Data

Laplante, Phillip A.
What every engineer should know about software engineering / Phillip A. Laplante.
p. cm. -- (What every engineer should know ; no. 1:40)
Includes bibliographical references and index.
ISBN-13: 978-0-8493-7228-5 (alk. paper)
ISBN-10: 0-8493-7228-3 (alk. paper)
1. Software engineering. I. Title. II. Series.

QA76.758.L327 2007
005.3--dc22 2006036497

Visit the Taylor & Francis Web site at
http://www.taylorandfrancis.com

and the CRC Press Web site at
http://www.crcpress.com

What Every Engineer Should Know: Series Statement

What every engineer should know amounts to a bewildering array of knowledge. Regardless of the areas of expertise, engineering intersects with all the fields that constitute modern enterprises. The engineer discovers soon after graduation that the range of subjects covered in the engineering curriculum omits many of the most important problems encountered in the line of daily practice—problems concerning new technology, business, law, and related technical fields.

With this series of concise, easy-to-understand volumes, every engineer now has within reach a compact set of primers on important subjects such as patents, contracts, software, business communication, management science, and risk analysis, as well as more specific topics such as embedded systems design. These are books that require only a lay knowledge to understand properly, and no engineer can afford to remain uniformed of the fields involved.

Introduction

What is the goal of this book?

This is a book about software engineering, but its purpose is not to enable you to leap into the role of a fully trained software engineer. That goal would be impossible to achieve solely with the reading of any book. Instead, the goal of this book is to help you better understand the nature of software engineering as a profession, as an engineering discipline, as a culture, and as an art form. And it is because of its ever morphing, multidimensional nature that non-software engineers have so much difficulty understanding the challenges software engineers must face.

Many practicing software engineers have little or no formal education in software engineering. While software engineering is a discipline in which practice and experience are important, it is rare that someone who has not studied software engineering will have the skills and knowledge needed to efficiently build industrial strength, reliable software systems. While these individuals may be perfectly capable of building working systems, unless a deliberate software engineering approach is followed, the cost of development will probably be higher than necessary, and the cost of maintaining the system will certainly be higher than it ought to be.

How is this book different from other software engineering books?

It is different from other software engineering books and from every other book I have written in that it is in Socratic form; that is, in the form of questions and answers. In some places I have shamelessly reused material from my books, particularly *Software Engineering for Image Processing Systems* (with attributions), but even then, significant rewriting was needed to place the material in the appropriate form of discourse. Indeed, in this present text I have generalized the concepts from that of predecessors to address the broader needs of all kinds of engineers.

Can this book convert me into a software engineer?

I don't promise that after reading this book you will become a master software engineer — no book can deliver on that promise. What this book will do, I hope, is help you better understand the limits of software engineering and the advances that have been made over the years. If nothing else, it is my hope that you will come away from reading this book with a deeper, more sympathetic understanding of the software engineer as an engineer.

Are software engineers really engineers?

Yes, the software engineer should be regarded as an engineer, particularly if he has the proper training, discipline, and mindset.

How should I use this book?

To get the most benefit from this book, I suggest you use it in one or more of the following ways:

- Read it through in its entirety to provide a general framework for and understanding of the profession of software engineering.
- Use it as a regular reference when questions about software, software engineering, or software engineers arise. You will find most of your questions directly addressed in this book.
- Skip around and read sections as needed to answer specific questions. There is no harm in reading this book out of order; after all, it was written out of order.

Who is the intended audience?

The target reader is the practicing engineer who has found he must write software, integrate off-the-shelf software into the systems he builds, or who works with software engineers on a regular basis. Undergraduate and graduate engineering students would be well served to have this book for reference, as it is likely that they will find themselves in the position of building software, and it is good to establish a rigorous framework early in their careers.

Did anyone help you with this book?

I have to acknowledge the help of several people along the way.

Some of this book is derived from lectures given by Drs. Colin Neill, Raghu Sangwan, and myself in the course, "Advanced Software Engineering," at The Pennsylvania State University (Penn State) School of Graduate Professional Studies. Some of the other material comes from my graduate software project management and software testing courses.

Drs. Sangwan and Neill and Professor Sally Richmond also reviewed various portions of the text and provided constructive criticism and useful ideas.

Dr. Pamela Vercellone-Smith offered some of the information on software evolution and reviewed various portions of the text.

My friend, Tom Costello, who is an expert in Open Source software, provided a great deal of information for and a review of Chapter 8.

Another friend, Will Gilreath, reviewed early drafts of the manuscript and provided many insights and some of the sample code.

Gary Birkmaier, my brother-in-law and principal software engineer with 25+ year's experience, reviewed and commented on the manuscript.

Chris Garrell, a former graduate student, provided the software requirements and design documents in Appendices A, B, and C. Chris is a registered professional engineer (civil) and also a holds a Master's degree in software engineering (the perfect combination for someone designing a wet well control system). He also reviewed and provided feedback on the finished manuscript.

Ken Sassa, another graduate student, provided the software archeology examples.

Over the years, many students in my graduate software engineering courses (many of them with degrees in various engineering disciplines) have contributed ideas that have influenced this book. I wish to thank them collectively.

I must also give thanks to my long-suffering wife and children for allowing me to work on this book when they would have preferred my undivided attention.

Finally, I would like to thank the wonderful folks at Taylor & Francis, particularly my editor, Allison Taub; Publisher, Nora Konopka; and my friends in the production department, particularly Helena Redshaw.

Are there copyrights and trademarks to be cited?

As noted previously, some of this book has been excerpted or adapted, with permission from the author's own text or others that are published by the Taylor & Francis Publishing Group. These are:

- *Dictionary of Computer Science, Engineering, and Technology,* Phillip A. Laplante (Editor), CRC Press, 2001.
- *Lightweight Enterprise Architectures,* Fenix Theuerkorn, Auerbach Publications, 2005.
- *Real Process Improvement Using CMMI,* Michael West, Auerbach Publications, 2004.
- *Software Engineering for Image Processing Systems,* Phillip A. Laplante, CRC Press, 2003.
- *Software Engineering Handbook,* Jessica Keyes, Auerbach Publications, 2003.

- *Software Engineering Measurement*, John C. Munson, Auerbach Publications, 2003.
- *Software Engineering Processes: Principles and Applications*, Yingxu Wang and Graham King, CRC Press, 2000.
- *Software Engineering Quality Practices*, Kurt Kandt, Auerbach Publications, 2005.
- *Software Testing and Continuous Quality Improvement*, 2nd ed., William E. Lewis, Auerbach Publications, 2005.
- *Software Testing: A Craftsman's Approach*, 2nd ed., Paul Jorgensen, CRC Press, 2002.
- *The Computer Science and Engineering Handbook*, Allen B. Tucker, Jr. (editor-in-chief), CRC Press, 1997.

Where more substantial portions of material have been used verbatim, or figures reproduced, it is so noted. Otherwise, these texts may be considered part of the general reference material for preparing this book.

Do you want to dedicate this book?

This book is dedicated to the many teachers, both academic and professional, who have helped me better understand software engineering over the last 25 years.

Can you tell me about yourself?

I am a professor of software engineering and a member of the graduate faculty at Penn State. I am also the Chief Technology Officer of the Eastern Technology Council, a nonprofit business advocacy group serving the Greater Philadelphia Metropolitan Area. Before joining Penn State, I was a professor, and later senior academic administrator, at several other colleges and universities.

Before my academic career, I spent almost eight years as a software engineer and project manager working on avionics (including the space shuttle), CAD, and software test systems. I have written or edited 22 books and more than 140 papers, articles, and editorials.

Over the years I have worked with, and for, many kinds of engineers. Non-software engineers have worked for me as well, and I have had the pleasure of teaching many hundreds of practicing engineers of various types about software engineering. This text, then, represents a compendium of what engineers should know about software engineering.

As for my "scholarly" credentials, I earned a B.S. and Ph.D. in computer science and an M.Eng. in electrical engineering from Stevens Institute of Technology, and an M.B.A. from the University of Colorado. In addition to my academic pursuits, I still consult regularly for the software industry, including Fortune 1000 companies and smaller software development houses.

Table of Contents

1

The Profession of Software Engineering

Outline

- Software engineering as an engineering profession
- Standards and certifications
- Misconceptions about software engineering

1.1 Introduction

If you want to start a debate among your engineering friends, ask the question, "Is software engineering real engineering?" Unfortunately, I suspect that if your friends are from one of the "hard" engineering disciplines such as mechanical, civil, chemical, and electrical, then their answers will be "no." This is unfortunate because software engineers have been trying for many years to elevate their profession to a level of respect granted to the hard engineering disciplines.

There are strong feelings around many aspects of the practice of software engineering — licensure, standards, minimum education, and so forth. Therefore, it is appropriate to start a book about software engineering by focusing on these fundamental issues.

1.2 Software Engineering as an Engineering Profession

What is software engineering?

Software engineering is "a systematic approach to the analysis, design, assessment, implementation, test, maintenance and reengineering of software, that is, the application of engineering to software. In the software engineering approach, several models for the software life cycle are defined, and many methodologies for the definition and assessment of the different phases of a life-cycle model" [Laplante 2001].

The profession of software engineering encompasses all aspects of conceiving, communicating, specifying, designing, building, testing, and maintaining software systems. Software engineering activities also include everything to do with the production of the artifacts related to software engineering such as documentation and tools.

There are many other ancillary activities to software engineering, one of which is the programming of the code or coding. But if you were stopped on the street by a pedestrian and asked to give a one-word definition for software engineering, your answer should be, "modeling." If you had two words to give, you might say, "modeling" and "optimization."

Modeling is a translation activity. The software product concept is translated into a requirements specification. The requirements are converted into a design. The design is then converted into code, which is automatically translated by compilers and assemblers, which produce machine executable code.

In each of these translation steps, however, errors are likely to be introduced either by the humans involved or by the tools they use. Thus, the practice of software engineering involves reducing translation errors through the application of correct principles.

The optimization part deals with finding the most economical translation possible. "Economical" means in terms of efficiency, clarity, and other desirable properties, which will be discussed later.

Is software engineering an engineering discipline?

The answer to this question depends on whom you ask. Many readers will argue that software engineering is not a true engineering discipline because there are no fundamental theorems grounded in the laws of physics (more on this later). Even some software engineering experts would add that there is still too much "art" in software engineering; that is, *ad hoc* approaches instead of rigorous ones. To further tarnish the image of software engineering, many self-styled practitioners do not have the appropriate background to engage in software engineering. These frauds help propagate the worst stereotypes by exemplifying what software engineering is not and should not be.

Perhaps the greatest assault on the reputation of software engineering and engineers occurs because of the eagerness to bring the software to the market. Of all the symptoms of poor software engineering, this is the one that management is most likely to condone.

Nevertheless, software engineering is trying to become a true engineering discipline through the development of more rigorous approaches, the evangelization of standards, the nurturing of an accepted body of knowledge for the profession, and proper education of software engineers.

What is the difference between software engineering and systems engineering?

There is a great deal of similarity in the activities conducted in software and systems engineering. Table 1.1, adapted from an excellent introduction to

TABLE 1.1

System Engineering Functions Correlated to Software System Engineering

System Engineering Function	Software Engineering Function	Software Engineering Description
Problem definition	Requirements analysis	Determine needs and constraints by analyzing system requirements allocated to software
Solution analysis	Software design	Determine ways to satisfy requirements and constraints, analyze possible solutions, and select the optimum one
Process planning	Process planning	Determine product development tasks, precedence, and potential risks to the project
Process control	Process control	Determine methods for controlling project and process, measure progress, and take corrective action where necessary
Product evaluation	Verification, validation, and testing	Evaluate final product and documentation

Source: Adapted from Thayer, R.H., Software system engineering: a tutorial, *Computer*, 35(4), 68–73, 2002.

software systems engineering by Richard Thayer, provides a summary of these activities. Take care when interpreting Table 1.1, as it has the tendency to suggest that the software engineering process is strictly a liner sequential (Waterfall) one. Various models of software development will be discussed shortly. Also note that there is no mention of "coding" in Table 1.1. This is not an inadvertent omission. In fact, the writing of code can be the least engineering-like activity that a software engineer can undertake.

What is the history of software engineering?

Although early work in software development and software engineering began in the late 1950s, many believe that software engineering first became a profession as a result of a NATO sponsored conference on software engineering in 1968. It is certainly true that this conference fueled a great deal of research in software engineering [Marciniak 1994].

Few people in the profession called themselves "software engineers" until mid- to late-1989, and major university programs in software engineering did not emerge until the late 1980s and early 1990s.

What is the role of the software engineer?

The production of software is a problem-solving activity that is accomplished by modeling. As a problem-solving, modeling discipline, software engineering is a human activity that is biased by previous experience, and is subject to human error. Therefore, the software engineer should recognize and try to eliminate these errors.

Software engineers should also strive to develop code that is built to be tested, that is designed for reuse, and that is ready for inevitable change. Anticipation of problems can only come from experience and from drawing upon a body of software practice experience that is more than 50 years old.

How do software engineers spend their time on the job?

Software engineers probably spend less than 10% of their time writing code. The other 90% of their time is involved with other activities that are more important than writing code. These activities include:

1. Eliciting requirements
2. Analyzing requirements
3. Writing software requirements documents
4. Building and analyzing prototypes
5. Developing software designs
6. Writing software design documents
7. Researching software engineering techniques or obtaining information about the application domain
8. Developing test strategies and test cases
9. Testing the software and recording the results
10. Isolating problems and solving them
11. Learning to use or installing and configuring new software and hardware tools
12. Writing documentation such as users manuals
13. Attending meetings with colleagues, customers, and supervisors
14. Archiving software or readying it for distribution

This is only a partial list of software engineering activities. These activities are not necessarily sequential and are not all encompassing. Finally, most of these activities can recur throughout the software life cycle and in each new minor or major software version. Many software engineers specialize in a small subset of these activities, for example, software testing.

What kind of education do software engineers need?

Ideally, software engineers will have an undergraduate degree in software engineering, computer science, or electrical engineering with a strong emphasis on software systems development. While it is true that computer science and software engineering programs are not the same, many computer science curricula incorporate significant courses on important aspects of software engineering. Unfortunately, there are not many undergraduate programs in software engineering. Most software engineering courses are taught under the auspicious of the computer science department.

An alternative path to the proper education in software engineering would be an undergraduate degree in a technical discipline and a Master's degree in software engineering (such as the degree that I am involved with at Penn State). Yet another path would be any undergraduate degree, significant experiential learning in software engineering on the job, and an appropriate Master's degree.

Another aspect of education involves the proper background in the domain area in which the software engineer is practicing. So, a software engineer building medical software systems would do well to have significant formal education in the health sciences, biology, medicine, and the like. One of my favorite students works in the medical informatics field; he has a Bachelor's degree in nursing, a Master's degree in software engineering, and significant experience — the perfect combination.

What kind of education do software engineers typically have?

Here is where the problem occurs. In my experience, many practicing software engineers have little or no formal education in software engineering. While software engineering is a discipline in which practice and experience are important, it is rare for someone who has not studied software engineering to have the skills and knowledge needed to efficiently and regularly build industrial strength, reliable software systems.

I know someone who is an "XYZ* Certified Engineer"; is he a software engineer?

He is definitely not a software engineer. There is an important distinction between certification and licensing. Certification is a voluntary process administered by a non-government entity. Licensing is a mandatory process controlled by a state licensing board. Some companies tend to avoid the use of the word "engineer" in the designation because the courts generally rule in favor of restricting the use of the term. One notable exception is Novell, which successfully defended its right to use "engineer" in its Certified Novell Engineer (CNE) designation in court decisions in Illinois in 1998 and Nevada in 2000 [IIE 2000].

Why are there so many software engineers without the proper education?

Shortages of trained software engineers in the 1980s and 1990s led to aggressive hiring of many without formal training in software engineering. This situation is commonly found in companies building engineering products where the "software engineers" were probably trained in some other technical discipline (for example, electrical engineering, physics, or mechanical engineering) but not in software engineering. In other cases, there is a tendency to move technicians into programming jobs as instrument interfaces move from hardware- to software-based. Finally, in some cases, long tenured employees, often without any technical experiences but with familiarity of the company's products and processes, move into software development. While all

* Where "XYZ" stands for some major company.

these persons may be perfectly capable of building working systems, unless a deliberate software engineering approach is followed, the cost of development and maintenance will be higher than necessary.

Can software engineering programs be accredited?

Yes, a body known as CSAB can accredit undergraduate software engineering programs. The acronym CSAB formerly stood for "Computing Sciences Accrediting Board," but is now used without elaboration. CSAB is a participating body of ABET (formerly known as the "Accreditation Board for Engineering and Technology" but also known only by its acronym). ABET accredits other kinds of undergraduate engineering programs. Within ABET, the Computing Accreditation Commission accredits programs in computer science and information systems, while the Engineering Accreditation Commission accredits programs in software engineering and computer engineering.

The relevant member societies of ABET for software engineering are the Association for Computing Machinery, the Institute of Electrical and Electronics Engineers (Computer Society), and the Association for Information Systems.

Is professional licensure available for software engineers?

At this writing, Texas is the only state that requires licensing of engineers who build systems involving "the application of mathematical, physical, or computer sciences to activities such as real-time and embedded systems, information or financial systems, user interfaces, and networks" [Statute 2006]. However, practitioners can sometimes avoid the rigorous licensing examination via a waiver rule that allows for recognition of significant experience (as little as 12 years).

In some cases, software engineers can obtain professional licensing in another engineering discipline (for example, I am licensed in the Commonwealth of Pennsylvania as an electrical engineer). However, this kind of licensure is only possible if the software engineer has the relevant qualifications for licensure in the alternative discipline. In addition, it is unclear what value the designation of "PE" holds for a software engineer — friends are impressed that I hold a PE license, but I have never seen a job advertisement for a practicing software engineer that required a PE license.

There are many proprietary certifications for software engineering practitioners. Are any of these valuable to a software engineer?

Not really. Certifications can be obtained in the technology *du jour* from such companies as Borland, Cisco, HP, IBM, Microsoft, and many others by passing tests. But the knowledge needed to pass these test has little to do with the discipline of software engineering. Instead, they often involve the memorization of rote steps needed for software installation and configuration.

What is the IEEE Computer Society CSDP certification?

The Institute of Electrical and Electronics Engineers' (IEEE) largest special interest group is the Computer Society, with over 80,000 members worldwide. The Computer Society, in conjunction with the Association for Computing Machinery (ACM), which also has over 80,000 members, has been leading the charge to professionalize software engineering. One of their initiatives is the Certified Software Development Professional (CSDP) certification.

Aimed at midlevel software engineers, the objectives of the CSDP program are to

"encourage self-assessment by offering guidelines for achievement

identify persons with acceptable knowledge of the principles and practices of software engineering

recognize those who have demonstrated a high level of competence in the profession

encourage continuing education

raise standards of the profession for the public at large"

"The CSDP is a comprehensive program that encourages individuals to draw from a broad base of software knowledge. The program is designed to measure the level of knowledge and competence that individuals have achieved in software engineering through experience, training, and education. CSDP is a professional certification program, but it is not licensure. CSDP is a credential of interest to many in the profession, particularly in safety, and mission-critical systems, but it is not for everyone" [CSDP 2006].

1.3 Standards and Certifications

Are there standards for software engineering practices, documentation, and so forth?

There are many, and I list them below for your reference. The title of the standard is self-explanatory. If the titles are not recognizable now, they will be after you have read this book.

IEEE Std 610.12-1990, IEEE Standard Glossary of Software Engineering Terminology

IEEE Std 1062, 1998 Edition, IEEE Recommended Practice for Software Acquisition

ISO/IEC 12207:1995 — Information Technology — Software Life-Cycle Processes

IEEE/EIA 12207 — US standard implementation of ISO/IEC std 12207:1995:

IEEE/EIA Std 12207.0-1996, Software Life Cycle Processes

IEEE/EIA Std 12207.1-1997, Software Life Cycle Processes — Life Cycle Data

IEEE/EIA Std 12207.2-1997, Software Life Cycle Processes — Implementation Considerations

IEEE Std 1228-1994, IEEE Standard for Software Safety Plans

Process Standards

IEEE Std 730-1998, IEEE Standard for Software Quality Assurance Plans

IEEE Std 730.1-1995, IEEE Guide for Software Quality Assurance Planning

IEEE Std 828-1998, IEEE Standard for Software Configuration Management Plans

IEEE Std 1008-1987, IEEE Standard for Software Unit Testing*

IEEE Std 1012-1998, IEEE Standard for Software Verification and Validation

IEEE Std 1012a-1998, Supplement to IEEE Standard for Software Verification and Validation: Content Map to IEEE/EIA 12207.1-1997

IEEE Std 1028-1997, IEEE Standard for Software Reviews

IEEE Std 1042-1987, IEEE Guide to Software Configuration Management**

IEEE Std 1045-1992, IEEE Standard for Software Productivity Metrics

IEEE Std 1058-1998, IEEE Standard for Software Project Management Plans

IEEE Std 1059-1993, IEEE Guide for Software Verification and Validation Plans

IEEE Std 1074-1997, IEEE Standard for Developing Software Life Cycle Processes

IEEE Std 1219-1998, IEEE Standard for Software Maintenance

Product Standards

IEEE Std 982.1-1988, IEEE Standard Dictionary of Measures to Produce Reliable Software

IEEE Std 982.2-1988, IEEE Guide for the Use of Standard Dictionary of Measures to Produce Reliable Software

IEEE Std 1061-1998, IEEE Standard for Software Quality Metrics Methodology

---

* Reaffirmed in 1993.

** Reaffirmed in 1993.

IEEE Std 1063-1987, IEEE Standard for Software User Documentation*

IEEE Std 1465-1998, IEEE Standard Adoption of International Standard ISO/IEC 12199:1994 (E) — Information Technology — Software packages — Quality requirements and testing

Resource and Technique Standards

IEEE Std 829-1998, IEEE Standard for Software Test Documentation

IEEE Std 830-1998, IEEE Recommended Practice for Software Requirements Documentation

IEEE Std 1016-1998, IEEE Recommended Practice for Software Design Descriptions

IEEE Std 1044-1993, IEEE Standard Classification for Software Anomalies

IEEE Std 1044.1-1995, IEEE Guide to Classification for Software Anomalies

IEEE Std 1348-1995, IEEE Recommended Practice for the Adoption of Computer-Aided Software Engineering (CASE) Tools

IEEE Std 1420.1-1995, IEEE Standard for Information Technology — Software Reuse — Data Model for Reuse Library Interoperability: Basic Interoperability Data Model (BIDM)

IEEE Std 1420.1a-1996, Supplement to the IEEE Standard for Information Technology — Software Reuse — Data Model for Reuse Library Interoperability: Asset Certification Framework

IEEE Std 1430-1996, IEEE Guide for Information Technology — Software Reuse — Concept of Operations for Interoperating Reuse Libraries

IEEE Std 1462-1998, IEEE Standard Adoption of ISO/IEC 14102:1995 — Information Technology — Guidelines for the Evaluation and Selection of CASE Tools

Of course there are many other standards issued by various organizations around the world covering various aspects of software engineering and computing sciences. The above selection is provided both for referencing purposes and to illustrate the depth and breadth of the software engineering standards that exist.

What is the Software Engineering Body of Knowledge?

The Software Engineering Body of Knowledge (abbreviated as SWEBOK but often pronounced as "sweebock") describes the "sum of knowledge within the profession of software engineering." Since 1993, the IEEE Computer Society and the ACM have been actively promoting software engineering as a profession, notably through their involvement in accreditation activities

* Reaffirmed in 1993.

described before and in the development of a software engineering body of knowledge [Bourque et al. 1999].

The objectives of the Guide to the SWEBOK project are to:

"characterize the contents of the Software Engineering Body of Knowledge;

provide a topical access to the Software Engineering Body of Knowledge;

promote a consistent view of software engineering worldwide;

clarify the place of, and set the boundary of, software engineering with respect to other disciplines such as computer science, project management, computer engineering and mathematics;

provide a foundation for curriculum development and individual certification and licensing material." [Bourque et al. 1999]

Knowledge areas include:

professionalism engineering economics

software requirements

software design

software construction and implementation

software testing

software maintenance

software configuration management

software engineering management

software engineering process

software engineering tools and methods

software quality

The "generally accepted knowledge" is described as follows:

Generally accepted — knowledge based on traditional practices that have been adopted by various organizations

Advanced and research — innovative practices tested and used only by some organizations and concepts still being developed and tested in research organizations

Specialize — practices used only for certain types of software [Bourque et al. 1999]

Software engineers must also be knowledgable in specifics of their particular application domain. For example, avionics software engineers need to have a great deal of knowledge of aerodynamics; software engineers for financial systems need to have knowledge in the banking domain, and so forth.

Are there any "fundamental theorems" of software engineering?

Software engineering has often been criticized for its lack of a rigorous, formalized approach. And in those cases where formalization is attempted, it is often perceived artificial and hence ignored (or accused of being computer science, not software engineering). Even a few of my most respected colleagues seem to hold this view. But there are some results in computer science, mathematics, and other disciplines that, while rigorous, can be shown to be applicable in a number of practical software engineering settings.

Rigor in software engineering requires the use of mathematical techniques. Formality is a higher form of rigor in which precise engineering approaches are used. In the case of the many kind of systems, such as real-time, formality further requires that there be an underlying algorithmic approach to the specification, design, coding, and documentation of the software. In every course I teach, I try to be rigorous. For example, I introduce finite state machines because students can readily see that they are practical yet formal.

It has been stated over and over again (without convincing proof) that high-tech jobs are leaving the U.S. partly because Americans are inadequately trained in mathematics as compared to other nationalities. Yet most people decry the need for mathematics in software engineering.

The most laudable efforts to justify the need for mathematics education in software engineering and computer science offer pedagogical arguments centering on the need to demonstrate higher reasoning and logical skills. While these arguments are valid, most students and critics will not be satisfied by them. Software engineering students (and computer science students) want to know why they must take calculus and discrete mathematics in their undergraduate programs because they often do not see uses for it.

Demming [2003] makes a plea for "great principles of computing;" that is, the design principles (simplicity, performance, reliability, evolvability, and security) and the mechanics (computation, communication, coordination, automation, and recollection). But perhaps there are no such grand theories, but rather many simple rules. Here is a list of some of my favorites:

Bayes' Theorem

Böhm-Jacopini Rule

Cantor's Diagonal Argument

Chebyshev's Inequality

Little's Law

McCabe's Cyclomatic Complexity Theorem

von Neumann's Min Max Theorem

Baye's Theorem provides the underpinning for a large body of artificial intelligence using Bayesian Estimation.

Böhm-Jacopini's Rule shows that all programs can be constructed using only a `goto` statement. This theory has important implications in computability and compiler theory, among other places.

Cantor's Diagonal Argument was used by mathematician Georg Cantor to show that that real numbers are uncountably infinite. But Cantor's Argument can also be used to show that the Halting Problem is undecidable; that is, there is no way to *a priori* prove that a computer program will stop under general conditions.

Chebyshev's Inequality for a random variable x with mean σ and standard deviation, is stated as

$$1 - \frac{1}{k^2} \leq P(|x - \mu| \geq k\sigma).$$

That is, the probability that the random variable will differ from its mean by k standard deviations is 1–1 over k^2. Chebyshev's Inequality can be used to make all kinds of statements about confidence intervals. So, for example, the probability that random variable x falls within two standard deviations of the mean is 75% and the probability that it falls within six deviations of the mean (six-sigma) is about 99.99%. This result has important implications in software quality engineering.

Little's Law is widely used in queuing theory as a measure of the average waiting time in a queue. Little's Law also has important implications in computer performance analysis.

McCabe's Cyclomatic Complexity Theorem demonstrates the maximum number of linearly independent code paths in a program, and is quite useful in testing theory and in the analysis of code evolution. These features are discussed later.

Finally, von Neumann's Min Max Theorem is used widely in economics and optimization theory. Min Max approaches can also be used in all kinds of software engineering optimization problems from model optimization to performance improvement.

Although there are other many mathematical concepts familiar to all engineers that could be introduced in my software engineering classes, the aforementioned ones can be easily shown to be connected to one or more very practical notions in software engineering. Still, it is true that the discipline of software engineering is lacking grand theory, such as Maxwell's Equations or the various Laws of Thermodynamics or even something as simple as the Ideal Gas Law in chemistry.

1.4 Misconceptions about Software Engineering

Why is software so buggy and unreliable?

It is unclear if software is more unreliable than any other complex engineering endeavor. While there are sensational examples of failed software, there are just as many examples of failed engineered structures, such as bridges collapsing, space shuttles exploding, nuclear reactors melting down, and so on.

To me, it seems that software gets a bad rap. Oftentimes when a project fails, software engineering is blamed, not the incompetence of the managers, inadequacy of the people on the project, or the lack of a clear goal.

In any case, you have to prove that software is more unreliable than any other kind of engineering system, and I have seen no compelling evidence to support that contention; everything I have seen or heard is anecdotal.

I write software as part of my job; does that make me a software engineer?

No! Anyone can call himself a software engineer if he writes code, but he is not necessarily practicing software engineering. To be a software engineer, you need more than a passing familiarity with most of the concepts of this book.

But isn't software system development primarily concerned with programming?

As mentioned before, 10% or less of the software engineer's time is spent writing code. Someone who spends the majority of his or her time generating code is more aptly called a "programmer." Just as wiring a circuit designed by an electrical engineer is not engineering, writing code designed by a software engineer is not an engineering activity.

Can't software tools and development methods solve most or all of the problems pertaining to software engineering?

This is a dangerous misconception. Tools, software or otherwise, are only as good as the wielder. Bad habits and flawed reasoning can just as easily be amplified by tools as corrected by them. While software engineering tools are essential and provide significant advantages, to rely on them to remedy process or engineering deficiencies is naïve.

Isn't software engineering productivity a function of system complexity?

While it is certainly the case that system complexity can degrade productivity, there are many other factors that affect productivity. Requirements stability, engineering skill, quality of management, and availability of resources are just a few of the factors that affect productivity.

Once software is delivered, isn't the job finished?

No. At the very least, some form of documentation of the end product as well as the process used needs to be written. More likely, the software product will now enter a maintenance mode after delivery in which it will experience many recurring life cycles as errors are detected and corrected and features are added.

Aren't errors an unavoidable side effect of software development?

While it is unreasonable to expect that all errors can be avoided (as in every discipline involving humans), good software engineering techniques can minimize the number of errors delivered to a customer. The attitude that errors are inevitable can be used to excuse sloppiness or complacency, whereas an approach to software engineering that is intended to detect every possible error, no matter how unrealistic this goal may be, will lead to a culture that encourages engineering rigor and high quality software.

1.5 Further Reading

Certified Software Development Professional Program (CDSP), IEEE Computer Society, http://www.computer.org/portal/site/ieeecs/menuitem. c5efb9b8ade90 96b8a9ca0108bcd45f3/index.jsp?&pName=ieeecs_ level1&path= ieeecs/education/certification&file=index.xml&xsl=generic.xsl&, accessed September 14, 2006.

Institute of Industrial Engineers (IIE), State board pays Novell in "engineer" title suit. *IIE Solutions*, 32(1), 10, 2000.

Demming, P., Great principles of computing, *Commun. ACM*, 46(11), 15–20, 2003.

Laplante, P.A. (Editor-in-Chief), *Comprehensive Dictionary of Computer Science, Engineering and Technology*, CRC Press, Boca Raton, FL, 2001.

Laplante, P.A., Professional licensing and the social transformation of software engineers, *Technol. Soc. Mag., IEEE* , 24(2), 40-45, 2005.

Marciniak, J. (Ed.), *Encyclopedia of Software Engineering*, Vol. 2, John Wiley & Sons, New York, 1994, 528–532.

Statute, Texas Board of Professional Engineers, http://www.tbpe.state.tx.us/, accessed August 11, 2006.

Bourque, P., Dupuis, R., Abran, A., Moore, J.W., and Tripp, L.L., The Guide to the Software Engineering Body of Knowledge," *IEEE Software*, 16, 35–44, 1999.

Thayer, R.H., Software system engineering: a tutorial, *Computer*, 35(4), 68–73, 2002.

Tripp, L.L., Benefits of certification, *Computer*, 35(6), 31–33, 2002.

2

Software Properties, Processes, and Standards

Outline

- Characteristics of software
- Software processes and methodologies
- Software standards

2.1 Introduction

To paraphrase Lord Kelvin, if you can't measure that which you are talking about, then you really don't know anything about it. In fact, one of the major problems with portraying software engineering as a true engineering discipline is the difficulty with which we have in characterizing various attributes, characteristics, or qualities of software in a measurable way. I begin this chapter, then, with the quantification of various attributes of software. Much of the information for this discussion has been adapted from the excellent section on software properties found in Tucker [1996].

Every software process is an abstraction, but the activities of the process need to be mapped to a life-cycle model. There is a variety of software life-cycle models, which are discussed in this chapter. While significantly more time is focused on the activities of the waterfall model, most of these activities also occur in other life-cycle models. Indeed, it can be argued that most other life-cycle models are refinements of the waterfall model.

I conclude the chapter by discussing some of the previously mentioned software standards that pertain to software qualities, life cycles, and processes.

The latter three sections of the chapter are largely derived and updated from my other software engineering text [Laplante 2004].

At this point, it is also convenient to mention that I will be using two major examples to illustrate many points throughout the text. One provides the software control for an airline baggage inspection system. Anyone who has flown recently will be sufficiently familiar with this application domain. In particular, I am only interested in the aspect of the baggage handling system that scans baggage as it moves down the conveyor system. The objective of the scan is to use x-ray imaging and some appropriate imaging algorithm to identify suspicious baggage and remove it from the conveyor using some mechanical reject mechanism.

The second example is a software system to control the actions of a wet well pumping system. The software requirements specification (SRS), software design, and object models for this software system are contained in Appendix A, Appendix B, and Appendix C, and will be explained throughout the rest of the text.

2.2 Characteristics of Software

How do software engineers characterize software?

Software can be characterized by any number of qualities. External qualities, such as usability and reliability, are visible to the user. Internal qualities are those that may not be necessarily visible to the user, but help the developers to achieve improvement in external qualities. For example, good requirements and design documentation might not be seen by the typical user, but these are necessary to achieve improvement in most of the external qualities. A specific distinction between whether a particular quality is external or internal is not often made because they are so closely tied. Moreover, the distinction is largely a function of the software itself and the kind of user involved.

What is "software reliability"?

Software reliability can be defined informally in a number of ways. For example, can the user "depend on" the software? Other loose characterizations of a reliable software system include:

- The system "stands the test of time."
- There is an absence of known catastrophic errors (those that disable or destroy the system).
- The system recovers "gracefully" from errors.
- The software is robust.

For engineering-type systems, other informal views of reliability might include the following:

- Downtime is below a certain threshold.
- The accuracy of the system is within a certain tolerance.
- Real-time performance requirements are met consistently.

How do you measure software reliability?

Software reliability can be defined in terms of statistical behavior; that is, the probability that the software will operate as expected over a specified time interval. These characterizations generally take the following approach. Let S be a software system and let T be the time of system failure. Then the reliability of S at time t, denoted $r(t)$, is the probability that T is greater than t; that is,

$$r(t) = P(T > t) \tag{2.1}$$

This is the probability that a software system will operate without failure for a specified period.

Thus, a system with reliability function $r(t) = 1$ will never fail. However, it is unrealistic to have such expectations. Instead, some reasonable goal should be set. For example, in the baggage inspection system, a reasonable standard of reliability might be that the failure probability be no more than 10^{-9} per hour. This represents a reliability function of $r(t) = (0.99999999)^t$ with t in hours. Note that as $t \rightarrow \infty$, $r(t) \rightarrow 0$.

What is a failure function?

Another way to characterize software reliability is in terms of a real-valued failure function. One failure function uses an exponential distribution where the abscissa is time and the ordinate represents the expected failure intensity at that time (Equation 2.2).

$$f(t) = \lambda e^{-\lambda t} \quad t \geq 0 \tag{2.2}$$

Here the failure intensity is initially high, as would be expected in new software as faults are detected during testing. However, the number of failures would be expected to decrease with time, presumably as failures are uncovered and repaired (Figure 2.1). The factor λ is a system-dependent parameter.

What is a "bathtub curve"?

The bathtub curve (see Figure 2.2) is often used to explain the failure function for physical components that wear out, electronics, and even biological systems.

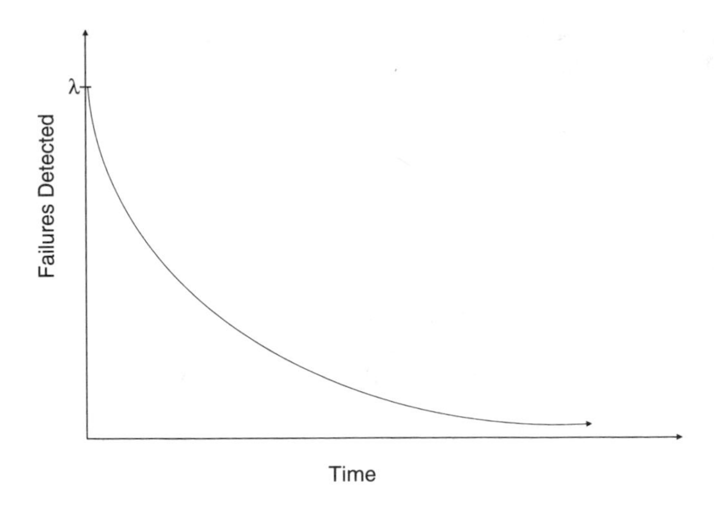

FIGURE 2.1
An exponential model of failure represented by the failure function $f(t) = \lambda e^{-\lambda t}$, $t \geq 0$. λ is a system-dependent parameter.

Obviously, we expect a large number of failures early in the life of a product (from manufacturing defects) and then a steady decline in failure incidents until later in the life of that product when it has "worn out" or, in the case of biological entities, died. But Brooks [1995] notes that the bathtub curve might also be useful in describing the number of errors found in a certain release of a software product.

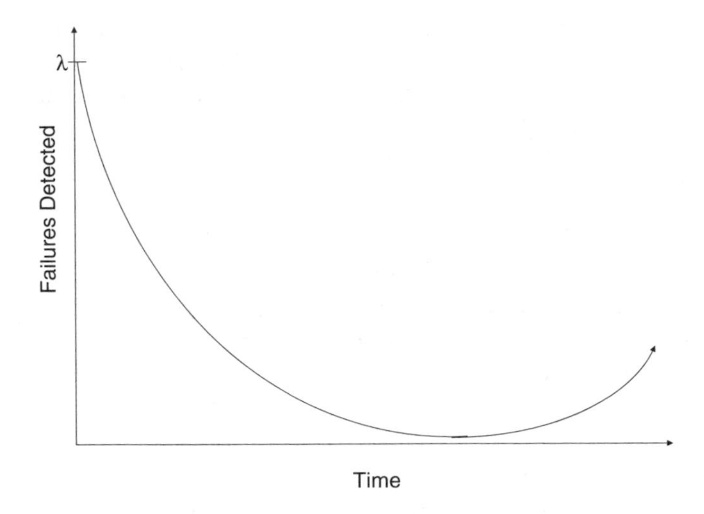

FIGURE 2.2
A software failure function represented by the bathtub curve.

But software doesn't wear out, so why would it conform to the bathtub curve?

It is clear that a large number of errors will be found in a particular software product early, followed by a gradual decline as defects are discovered and corrected, just as in the exponential model of failure. But we have to explain the increase in failure intensity later in time. There are at least three possible explanations. The first is that the errors are due to the effects of patching the software for bug fixes or new features.

The second reason for a late surge in failures is that the underlying hardware or operating system may have recently changed in a way that the software engineers did not anticipate.

Finally, additional failures could appear because of the increased stress on the software by expert users. That is, as users master the software and begin to expose and strain advanced features, it is possible that certain poorly tested functionality of the software is beginning to be used.

Can the traditional quality measures of MTFF or MTBF be used to stipulate reliability in the software requirements specification?

Yes, mean time to first failure (MTFF) or mean time between failures (MTBF) can be used. This approach to failure definition places great importance on the effective elicitation (discovery) and specification of functional requirements because the requirements define the software failure.

What is meant by the "correctness" of software?

Software correctness is closely related to reliability and the terms are often used interchangeably. The main difference is that minor deviation from the requirements is strictly considered a failure and hence means the software is incorrect. However, a system may still be deemed reliable if only minor deviations from the requirements are experienced. Correctness can be measured in terms of the number of failures detected over time.

What is software "performance"?

Performance is a measure of some required behavior — often with respect to some relative time constraint. For example, the baggage inspection system may be required to process 100 pieces of luggage per minute. But a photo reproduction system might be required to digitize, clean, and output color copies at a rate of one every two seconds.

How is software performance measured?

One method of measuring performance is based on mathematical or algorithmic complexity. Another approach involves directly timing the behavior of the completed system with logic analyzers and similar tools.

How do we characterize software usability?

Usability is a measure of how easy the software is for humans to use. Software usability is synonymous with ease-of-use, or user-friendliness.

Properties that make an application user-friendly to novice users are often different from those desired by expert users or software designers. Use of prototyping can increase the usability of a software system because, for example, interfaces can be built and tested by the user.

How do you measure software usability?

This quality is an elusive one. Usually informal feedback from users is used, as well as surveys, focus groups, and problem reports. Designer as apprentice, a requirements discovery and refinement technique that will be discussed in Chapter 3, can also be used to determine usability.

What is interoperability?

This quality refers to the ability of the software system to coexist and cooperate with other systems. For example, in embedded systems* the software must be able to communicate with various devices using standard bus structures and protocols.

In many systems, special software called middleware is written to enhance interoperability. In other cases, standards are used to achieve better interoperability.

How is interoperability measured?

Interoperability can be measured in terms of compliance with open system standards. These standards are typically specific to the application domain. For example, in the railway industry, the prevailing standard of interoperability is IEEE 1473 – 1999 [IEEE 1999].

What is an open system?

An open system is an extensible collection of independently written applications that cooperate to function as an integrated system. This concept is related to interoperability. Open systems differ from open source code, which is source code that is made available to the user community for improvement and correction. Open source code systems will be discussed in detail in Chapter 7.

What are the advantages of an open system?

An open system allows the addition of new functionality by independent organizations through the use of interfaces whose characteristics are published. Any software engineer can then take advantage of these interfaces, and thereby create software that can communicate using the interface. Open systems also permit different applications written by different organizations to interoperate.

---

* Embedded systems interact closely with specialized hardware in a unique environment. Both the baggage inspection system and wet well control system are embedded.

What is software "maintainability, evolvability, and repairability"?

Anticipation of change is a general principle that should guide the software engineer. A software system in which changes are relatively easy to make has a high level of maintainability. In the long run, design for change will significantly lower software life-cycle costs and lead to an enhanced reputation for the software engineer, the software product, and the company.

Maintainability can be decomposed into two contributing properties — evolvability and repairability. Evolvability is a measure of how easily the system can be changed to accommodate new features or modification of existing features. Repairability is the ability of a software defect to be easily repaired.

How do you measure maintainability, evolvability, and reparability?

Measuring these qualities is not always easy, and is often based on anecdotal observation. This means that changes and the cost of making them are tracked over time. Collecting this data has a twofold purpose. First, the costs of maintenance can be compared to other similar systems for benchmarking and project management purposes. Second, the information can provide experiential learning that will help to improve the overall software production process and the skills of the software engineers.

What is meant by "portability"?

Software is portable if it can run easily in different environments. The term environment refers to the hardware on which the system resides, the operating system, or other software in which the system is expected to interact.

The Java programming language, for example, was invented to provide a program execution environment that supported full portability across a wide range of embedded systems platforms and applications (see Chapter 5).

How is portability measured?

Portability is difficult to measure, other than through anecdotal observation. Person months required to perform the port is the standard measure of this property.

How do you make software more portable?

Portability is achieved through a deliberate design strategy in which hardware-dependent code is confined to the fewest code units as possible. This strategy can be achieved using either object-oriented or procedural programming languages and through object-oriented or structured approaches. All of these will be discussed later.

What is "verifiability"?

A software system is verifiable if its properties, including all of those previously introduced, can be verified easily.

How can you increase software verifiability?

One common technique for increasing verifiability is through the insertion of software code that is intended to monitor various qualities such as performance or correctness. Modular design, rigorous software engineering practices, and the effective use of an appropriate programming language can also contribute to verifiability.

What is "traceability" in software systems?

Traceability is concerned with the relationships between requirements, their sources, and the system design. Regardless of the process model, documentation and code traceability is paramount. A high level of traceability ensures that the software requirements flow down through the design and code and then can be traced back up at every stage of the process. This would ensure, for example, that a coding decision can be traced back to a design decision to satisfy a corresponding requirement.

Traceability is particularly important in embedded systems because often design and coding decisions are made to satisfy hardware constraints that may not be easily associated with a requirement. Failure to provide a traceable path from such decisions through the requirements can lead to difficulties in extending and maintaining the system.

Generally, traceability can be obtained by providing links between all documentation and the software code. In particular, there should be links:

- from requirements to stakeholders who proposed these requirements
- between dependent requirements
- from the requirements to the design
- from the design to the relevant code segments
- from requirements to the test plan
- from the test plan to test cases.

Are there other software qualities?

Martin [2002] describes such a set of software code qualities in the negative. That is, these are qualities of the code that need to be reduced or avoided altogether. They include:

Fragility — When changes cause the system to break in places that have no conceptual relationship to the part that was changed. This is a sign of poor design.

Immobility — When the code is hard to reuse.

Needless complexity — When the design is more elaborate than it needs to be. This is sometimes also called "gold plating."

Needless repetition — This occurs when cut-and-paste (of code) is used too frequently.

TABLE 2.1

Negative Code Qualities and Their Positives

Negative Code Quality	Positive Code Quality
Fragility	Robustness
Immobility	Reusability
Needless complexity	Simplicity
Needless repetition	Parsimony
Opacity	Clarity
Rigidity	Flexibility
Viscosity	Fluidity

Source: Martin, R.C., *Agile Software Development, Principles, Patterns, and Practices,* Prentice-Hall, Englewood Cliffs, NJ, 2002.

Opacity — When the code is not clear.

Rigidity — When the design is hard to change because every time you change something, there are many other changes needed to other parts of the system.

Viscosity — When it is easier to do the wrong thing, such as a quick and dirty fix, than the right thing.

The desirable opposites of these qualities are given in Table 2.1.

Achieving these qualities is a direct result of a good software architecture, solid software design, and effective coding practices, which are discussed in Chapters 3, 4, and 5, respectively.

Aren't there other software qualities that you left out?

Of course, there are many software qualities that could be discussed, some mainstream, others more esoteric or application-specific. For brevity, I have confined the discourse to the most commonly discussed qualities of software. In-depth discussion of these and other qualities to be considered can be found throughout the literature, for example, [Tucker, 1996].

2.3 Software Processes and Methodologies

What is a software process?

A software process is a model that describes an approach to the production and evolution of software. Software process models are frequently called "life-cycle" models, and the terms are interchangeable.

Isn't every software process model just an abstraction?

As with any model, a process model is an abstraction. But in this case, the model depicts the process of translation — from system concept, to requirements specification, to a design, then code, then finally via compilation and assembly, to the stored program form.

What benefits are there to using a software process model?

A good process model will help minimize the problems associated with each translation. A software process also provides for a common software development framework both within a project and across projects. The process allows for productivity improvements and it provides for a common culture, a common language, and common skills among organizational members. These benefits foster a high level of traceability and efficient communication throughout the project. In fact, it is very difficult to apply correct project management principles when an appropriate process model is not in place.

What is a software methodology?

The methodology describes the "how." It identifies how to perform activities for each period, how to represent the activities and products,* and how to generate products.

Aren't software process models and methodologies the same?

A software methodology is not the same as a software process. A software process is, in essence, the "what" of the software product life cycle. The process identifies and determines the order of phases within the life cycle. It establishes phase transition criteria, and indicates "what" is to be done in each phase and when to stop. However, the terms for process model and methodology are often used interchangeably (and, possibly, incorrectly). For example, there is both an agile software process model and many different agile methodologies, to be discussed shortly.

What is the waterfall life cycle model?

The terms waterfall, conventional, or linear sequential are used to describe a sequential model of nonoverlapping and distinctive activities related to software development. Collectively, the periods in which these activities occur are often referred to as phases or stages. While simplistic and dating back at least 30 years, the waterfall model is still popular. One survey, for example, showed that 35% of companies still use a waterfall model [Neill 2003].

How many phases should the waterfall model have?

The number of phases differs between variants of the model. As an example of a waterfall model, consider a software development effort with activity

* The term *artifact* is sometimes used to mean software or software-related products such as documentation.

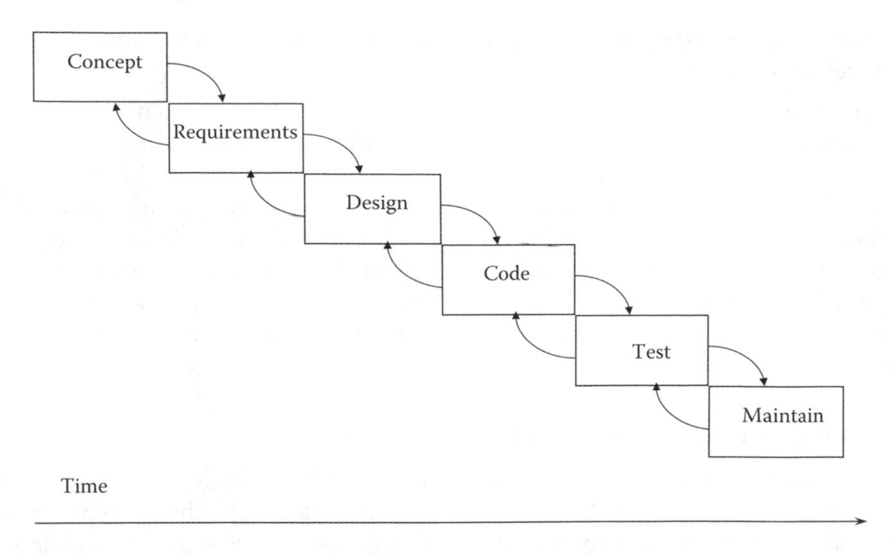

FIGURE 2.3
A waterfall life-cycle model. The forward arcs represent time sequential activities. The reverse arcs represent backtracking.

periods that occur in the following sequence:

Concept definition
Requirements specification
Design specification
Code development
Testing
Maintenance

The waterfall representation of this sequence is shown in Figure 2.3. Table 2.2 summarizes the activities in each period and the main artifacts of these activities.

TABLE 2.2
Phases of a Waterfall Software Life Cycle with Associated Activities and Artifacts

Phase	Activity	Output
Concept	Define project goals	White paper
Requirements	Decide what the software must do	Software requirements specification
Design	Show how the software will meet the requirements	Software design description
Development	Build the system	Program code
Test	Demonstrate requirements satisfaction	Test reports
Maintenance	Maintain system	Change requests, reports

What happens during the software conception phase of the waterfall process model?

The software conception activities include determination of the software project's needs and overall goals. These activities are driven by management directives, customer input, technology changes, and marketing decisions.

At the onset of the phase, no formal requirements are written, generally no decisions about hardware/software environments are made, and budgets and schedules cannot be set. In other words, only the features of the software product and possibly the feasibility of testing them are discussed. Usually, no documentation other than internal feasibility studies, white papers, or memos are generated.

Does the software conception phase really happen?

Some variants of the waterfall model do not explicitly include a conception period because the activity was either incorporated into the requirements definition or not thought to be part of the software project at all. Nonetheless, the concept activity does occur in every software product, even if it is implicit.

What happens during the requirements specification phase of the waterfall process model?

The main activity of this phase is creating the SRS. This activity is discussed in detail in Chapter 3.

Do any test activities occur during the requirements specification phase?

During this phase test requirements are determined and committed to a formal test plan. The test plan is used as the blueprint for the creation of test cases used in the testing phase, which is discussed later in the text.

The requirements specification phase can and often does occur in parallel with product conception and, as mentioned before, they are often not treated as distinct. It can be argued that the two are separate, however, because the requirements generated during conceptualization are not binding, whereas those determined in the requirements specification phase are (or should be) binding. This rather subtle difference is important from a testing perspective because the SRS represents a binding contract and, hence, the criteria for product acceptance. Conversely, ideas generated during system conceptualization may change and, therefore, are not yet binding.

What happens during the software design phase of the waterfall process model?

The main activity of software design is to develop a coherent, well-organized representation of the software system suitable to guide software development. In essence, the design maps the "what" from the SRS to the "how" of the software design description. Techniques for software design are discussed in Chapter 3.

Do any test activities occur during the software design phase?

Certain test activities occur concurrently with the preparation of the software design description. These include the development of a set of test cases based on the test plan. Techniques for developing test cases are discussed later.

Often during the software design phase problems in the SRS are identified. These problems may include conflicts, redundancies, or requirements that cannot be met with current technology. In real-time systems* the most typical problems are related to deadline satisfaction.

Usually, problems such as these require changes to the SRS or the granting of exemptions from the requirements in question. In any case, the problem resolution shows up as a specific directive in the software design description.

What happens during the software development phase of the waterfall process model?

This phase involves the production of the software code based on the design using best practices. These best practices will be discussed in Chapter 4.

What test activities occur during the software development phase?

During this phase the test team can build the test cases specified in the design phase in some automated form. This approach guarantees the efficacy of the tests and facilitates repeat testing.

When does the software development phase end?

The software development phase ends when all software units have been coded, unit tested, and integrated, and have passed the integration testing specified by the software designers.

What happens during the testing phase of the waterfall process model?

Although ongoing testing is an implicit part of the waterfall model, the model also includes an explicit testing phase. These testing activities (often called acceptance testing to differentiate them from code unit testing) begin when the software development phase has ended. During the explicit testing phase, the software is confronted with a set of test cases (module and system level) developed in parallel with the software and documented in a software test requirements specification (STRS). Acceptance or rejection of the software is based on whether it meets the requirements defined in the SRS using tests and criteria set forth in the STRS.

---

* Many software systems with which engineers deal are real-time systems; that is, performance satisfaction is based on both the correctness of the outputs, as well as the timeliness of those outputs. Hard real-time systems are those in which missing even a single deadline will lead to total system failure. Firm real-time systems can tolerate a few missed deadlines, while in soft real-time systems, missed deadlines generally lead to performance degradation only.

When does the testing phase end?

The testing phase ends when either the criteria established in the STRS are satisfied, or failure to meet the criteria forces requirements modification, design alteration, or code repair. Regardless of the outcome, one or more test reports are prepared which summarize the conduct and results of the testing. More on testing, including test stoppage criteria, will be discussed in Chapter 5.

What happens during the software maintenance phase of the waterfall process model?

The software maintenance phase activities generally consist of a series of reengineering processes to prolong the life of the system. Maintenance activities can be adaptive, which result from external changes to which the system must respond, corrective, which involves maintenance to correct errors, or perfective, which is all other maintenance including user enhancements, documentation changes, efficiency improvements, and so on. The maintenance activity ends only when the product is no longer supported.

In some cases, the maintenance phase is not incorporated into the life-cycle model, but instead treated as a series of new software products, each with its own waterfall life cycle.

How are maintenance corrections supposed to be handled?

Maintenance corrections are usually handled by making a software change and then performing regression testing. Another approach is to collect a set of changes and then regression test against the last set of changes.

The waterfall model looks artificial. Is there no backtracking?

Yes, as shown in Figure 2.3, backtracking transitions do occur. For example, new features, lack of sufficient technology, or other factors force reconsideration of the system purpose. Redesign may result in a return to the requirements phase during design. Similarly, a transition from the programming phase back to the design phase might occur due to a feature that cannot be implemented or caused by an undesirable performance result. This in turn may necessitate redesign, new requirements, or elimination of the feature. Finally, a transition from the testing phase to the programming or design phases may occur due to an error detected during testing. Depending on the severity of the error, the solution may require reprogramming, redesign, modification of requirements, or reconsideration of the system goals.

What is the V model for software?

The V model is a variant of the waterfall model. It represents a tacit recognition that there are testing activities occuring throughout the waterfall

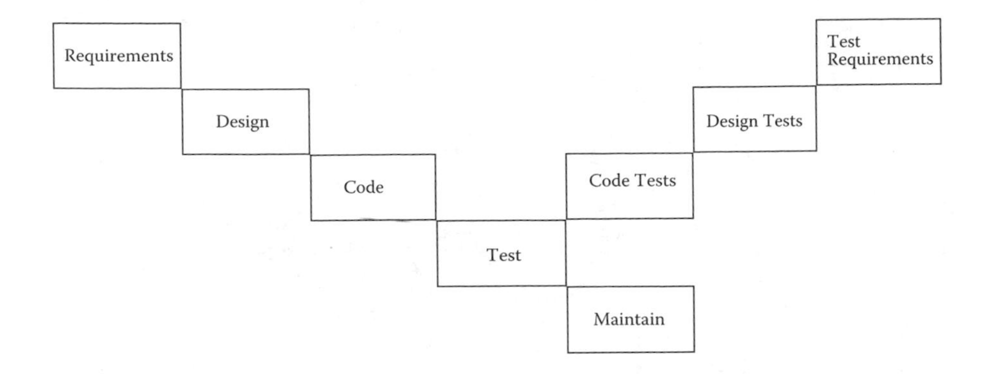

FIGURE 2.4
A V model for the software project life cycle. The concept phase is combined with the requirements phase in this instance.

software life cycle model and not just during the software testing period. These concurrent activities, depicted in Figure 2.4, are described alongside the activities occurring in the waterfall model.

For example, during requirements specification, the requirements are evaluated for testability and an STRS may be written. This document would describe the strategy necessary for testing the requirements. Similarly during the design phase, a corresponding design of test cases is performed. While the software is coded and unit tested, the test cases are developed and automated. The test life cycle converges with the software development life cycle during acceptance testing.

The point, of course, is that testing is a full life-cycle activity and that it is important to constantly consider the testability of any software requirement and to design to allow for such testability.

What is the spiral model for software?

The spiral model, suggested by Boehm [1988], recognizes that the waterfall model is not a realistic representation, nor is it necessarily a healthy one. Instead, the spiral model augments the waterfall model with a series of strategic prototyping and risk assessment activities throughout the life cycle. The spiral model is depicted in Figure 2.5.

Starting at the center of the figure, the product life cycle continues in a spiral path from the concept and requirements phases. Prototyping and risk analysis are used along the way to evaluate the feasibility of potential features. The added risk protection benefit from the extensive prototyping can be costly, but is well worth it, particularly in embedded systems. More will be mentioned about risk in Chapter 6.

More prototyping is used after a software development plan is written, and again after the design and tests have been developed. After that, the model behaves somewhat like a waterfall model.

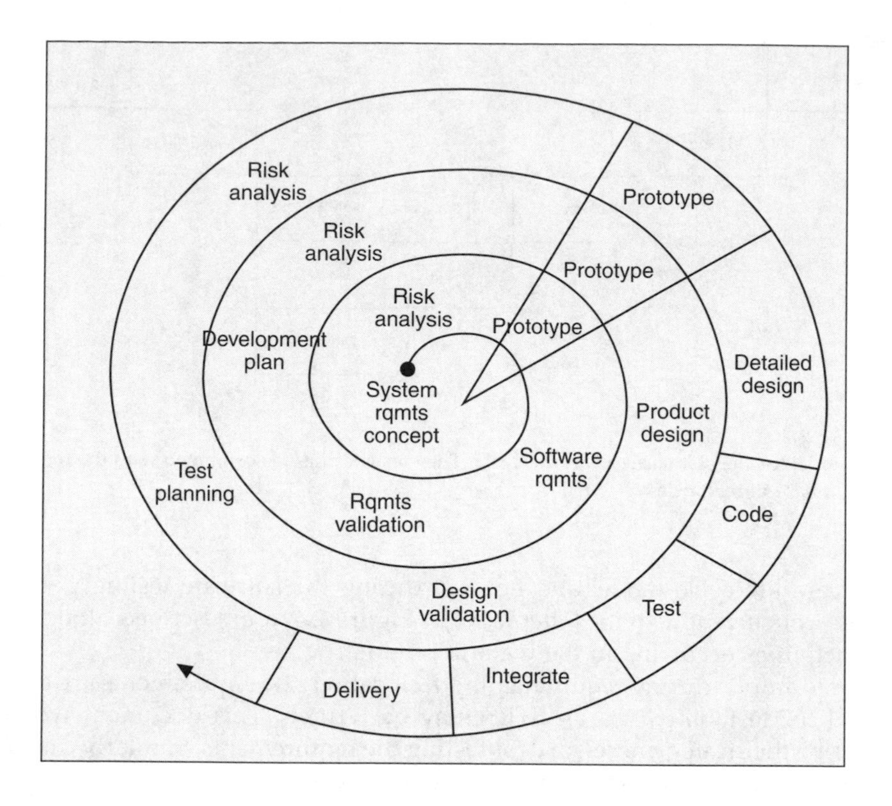

FIGURE 2.5
The spiral software model. Adapted from Boehm, B.W., A spiral model of software development and enhancement, *Computer*, 21(5), 61–72, 1988.

Is the spiral model widely used?

Apparently not; the aforementioned survey indicated that about 9% of organizations used it [Neill 2003].

What are evolutionary models?

Evolutionary life-cycle models promote software development by continuously defining requirements for new system increments based on experience from the previous version. Evolutionary models go by various names such as Evolutionary Prototyping, Rapid Delivery, Evolutionary Delivery Cycle, and Rapid Application Delivery (RAD).

In the evolutionary model, each iteration follows the waterfall model in that there are requirements, software design and testing phases. After the final evolutionary step, the system enters the maintenance phase, although it can evolve again through the conventional flow, if necessary.

The evolutionary model can be used in conjunction with embedded systems, particularly in working with prototype or novel hardware that come from simulators during development. Indeed, there may be significant benefits to this approach. First, early delivery of portions of the system can be generated, even though some of the requirements are not finalized. Then these early releases are used as tools for requirements elicitation, including timing requirements.

From the developers' point of view, those requirements that are clear at the beginning of the project drive the initial increment, but the requirements become clearer with each increment.

Are evolutionary models widely used?

Evolutionary models are gaining in popularity. For example, in the previously mentioned survey, almost 20% of respondents indicated its adoption [Neill 2003].

Are there any downsides to using evolutionary models?

Yes. For example, there may be difficulties in estimating costs and schedule when the scope and requirements are ill-defined. In addition, the overall project completion time may be greater than if the scope and requirements are established completely before design. Unfortunately, time apparently gained on the front end of a project because of early releases may be lost later because of the need for rework resulting from evolving requirements. Indeed, care must be taken to ensure that the evolving system architecture is both efficient and maintainable so that the completed system does not resemble a patchwork of afterthought add-ons. Finally, additional time must also be planned for integration and regression testing as increments are developed and added to the system. Some of the difficulties in using this approach in engineering systems can be mitigated, however, if the high-level requirements and overall architecture are established before entering an evolutionary cycle.

What is the incremental software model?

The incremental model is characterized by a series of detailed system increments, each increment incorporating new or improved functionality to the system. These increments may be built serially or in parallel depending on the nature of the dependencies among releases and on availability of resources.

What is the difference between incremental and evolutionary models?

The difference is that the incremental model allows for parallel increments. In addition, the serial releases of the incremental model are planned whereas in the evolutionary model, each sequential release is a function of the experience from the previous iteration.

Why use the incremental model?

There are several advantages to using the incremental model. These include ease of understanding each increment because of the decreased functionality, the use of successive increments in requirements elicitation, early development of initial functionality (which may aid in developing the real-time scheduling structure and for debugging prototype hardware), and successive building of operational functionality over time. The thinking is that software released in increments over time is more likely to satisfy changing user requirements than if the system were planned as a single overall release at the end of the same period. Finally, because the sub-projects are smaller, project management is more manageable for each increment.

Are there any downsides to the incremental model?

As with the evolutionary model, there may be increased system development costs as well as difficulties in developing temporal behavior and meeting timing constraints with a partially implemented system.

Is the incremental model used very much?

There is some evidence that the incremental model is fairly widespread. The aforementioned survey suggests that slightly more than 20% of companies are using it [Neill 2003].

What is the unified process model?

The unified process model (UPM) uses an object-oriented approach by modeling a family of related software processes using the unified modeling language (UML) as a notation. Like UML, UPM is a metamodel for defining processes and their components.

The UPM consists of four phases, which undergo one or more iterations. In each iteration some technical capability (software version or build) is produced and demonstrated against a set of criteria. The four phases in the UPM model are:

1. Inception: Establish software scope, use cases, candidate architecture, risk assessment.
2. Elaboration: Produce baseline vision, baseline architecture, select components.
3. Construction: Conduct component development, resource management and control.
4. Transition: Perform integration of components, deployment engineering, and acceptance testing.

Several commercial and open source tools support the UPM and provide a basis for process authoring and customization.

Where is the UPM used?

The UPM was developed to support the definition of software development processes specifically including those processes that involve or mandate the use of UML, such as the Rational Unified Process, and is closely associated with the development of systems using object-oriented techniques.

What are agile methodologies?

Agile software development methodologies are a subset of iterative methods* that focus on embracing change, and stress collaboration and early product delivery while maintaining quality. Working code is considered the true artifact of the development process. Models, plans, and documentation are important and have their value, but exist only to support the development of working software, in contrast with the other approaches already discussed. However, this does not mean that an agile development approach is a free-for-all. There are very clear practices and principles that agile methodologists must embrace.

Agile methods are adaptive rather than predictive. This approach differs significantly from those models previously discussed that emphasize planning the software in great detail over a long period of time and for which significant changes in the SRS can be problematic. Agile methods are a response to the common problem of constantly changing requirements that can bog down the more "ceremonial" upfront design approaches, which focus heavily on documentation at the start.

Agile methods are also "people-oriented" rather than process-oriented. This means they explicitly make a point of trying to make development "fun." Presumably, this is because writing SRSs and software design descriptions is onerous and, hence, to be minimized.

What are some agile methodologies?

Agile methodologies sometimes go by funny names like Crystal, Extreme Programming (XP), and Scrum. Other agile methods include dynamic systems development method (DSDM), feature-driven development, adaptive programming, and many more. We will look more closely at two of these, XP and Scrum.

What is Extreme Programming?

Extreme Programming** (XP) is one of the most widely used agile methodologies. XP is traditionally targeted toward smaller development teams and requires relatively few detailed artifacts. XP takes an iterative approach to its development cycles. However, whereas an evolutionary or iterative method may still have distinct requirements analysis, design, implementation, and testing phases similar to the waterfall method, XP turns these activities on their sides (Figure 2.6).

There is an initial analysis period to get things going, but after that, all activities occur more or less continuously throughout the development life cycle.

---

* Most people define agile methodologies as being incremental. But incremental development implies that the features and schedule of each delivered version are planned. In my experience, agile methodologies tend to lead to versions with feature sets and delivery dates that are almost always not as planned.

** Extreme programming is sometimes also written as "eXtreme Programming" to highlight the "XP."

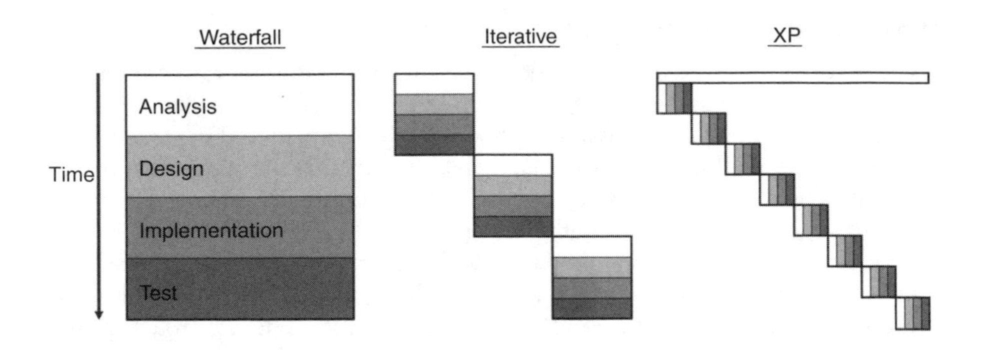

FIGURE 2.6
Comparison of waterfall, iterative, and XP development cycles. (From Beck, K., Embracing change with extreme programming, *Computer*, 32(10), 70–77, 1999.)

Some of my students use XP in software development classes and at work, and they widely report that it is effective, fun, and easy to use.

What are some of the practices of XP?

XP promotes a set of 12 core practices that help developers respond to and embrace inevitable change. The practices, which occur throughout the life of the software project, can be grouped according to four areas that cover planning, coding, designing, and testing (see Figure 2.7).

Planning
- User stories are written.
- Release planning creates the schedule.
- Make frequent small releases.
- The Project Velocity is measured.
- The project is divided into iterations.
- Iteration planning starts each iteration.
- Move people around.
- A stand-up meeting starts each day.
- Fix XP when it breaks.

Designing
- Simplicity.
- Choose a system metaphor.
- Use Class, Responsibilities, and Collaboration (CRC) cards for design sessions.
- Create spike solutions to reduce risk.
- No functionality is added early.
- Refactor whenever and wherever possible.

Coding
- The customer is always available.
- Code must be written to agreed standards.
- Code the unit test first.
- All production code is pair programmed.
- Only one pair integrates code at a time.
- Integrate often.
- Use collective code ownership.
- Leave optimization till last.
- No overtime.

Testing
- All code must have unit tests.
- Al code must pass all unit tests before it can be released.
- When a bug is found tests are created.
- Acceptance tests are run often and the score is published.

FIGURE 2.7
The rules and practices of XP. (From "The Rules and Practices of Extreme Programming," www.extremeprogramming.org/rules.html, accessed September 14, 2006.)

Some of the distinctive planning features of XP include holding daily stand-up meetings, making frequent small releases, and moving people around. Coding practices include having the customer constantly available, coding the unit test cases first, and employing pair-programming (a unique coding strategy where two developers work on the same code together). Removal of the territorial ownership of any code unit is another feature of XP.

Design practices include looking for the simplest solutions first, avoiding too much planning for future growth (speculative generality), and refactoring the code (improving its structure) continuously.

Testing practices include creating new test cases whenever a bug is found and unit testing for all code, possibly using frameworks such as XUnit, which is discussed in Chapter 6.

Can you say more about Scrum?

Scrum, which is named after a particularly contentious point in a rugby match, enables self-organizing teams by encouraging verbal communication across all team members and across all stakeholders. The fundamental principle of Scrum is that traditional problem definition solution approaches do not always work, and that a formalized discovery process is sometimes needed.

Scrum features a dynamic list of prioritized work to be done. Completion of a largely fixed set of backlogged items occurs in a series of short (approximately 30 days) iterations or sprints.

Each day, a brief meeting or Scrum is held in which progress is explained, upcoming work is described, and impediments are raised. A brief planning session occurs at the start of each sprint to define the backlog items to be completed. A brief postmortem or heartbeat retrospective occurs at the end of the sprint.

A "ScrumMaster" removes obstacles or impediments to each sprint. The ScrumMaster is not the leader of the team (as they are self-organizing) but acts as a productivity buffer between the team and any destabilizing influences.

Several major corporations have adopted Scrum with notable success. Some of my students also use Scrum in courses and it has proven to be effective.

Is there a case to be made for not using agile methods?

Like all engineering solution approaches, there are situations when agile software engineering should be used and there are situations when it should not be used. But it is not always easy to make this distinction. Ongoing misuse or misunderstanding can cloud the decision as to when to use agile approaches.

When should agile methodologies be used?

Boehm and Turner [2003] suggest that the way to assess whether agile methodologies should be used is to look at the project along a continuum of five dimensions:

1. size (in terms of number of personnel involved)
2. system criticality

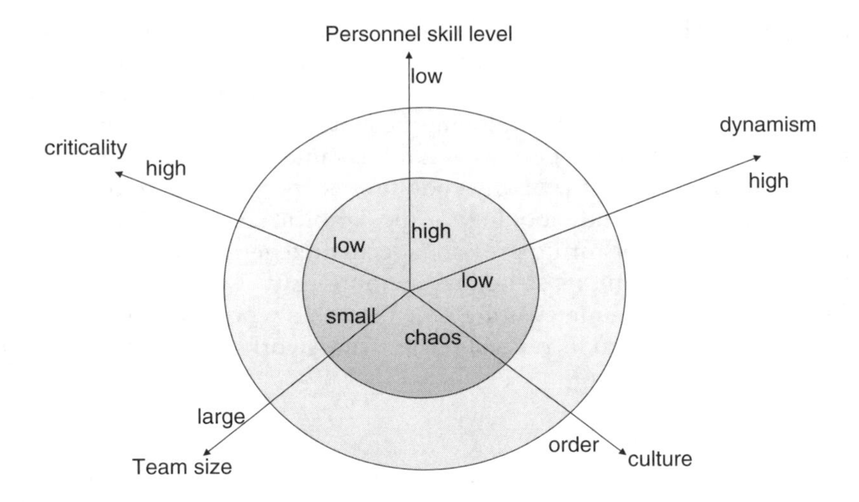

FIGURE 2.8
Balancing agility and discipline. (Adapted from Boehm, B.W. and Turner, R., *Balancing Agility and Discipline: A Guide to the Perplexed*, Addison-Wesley, Boston, MA, 2004.)

3. personnel skill level
4. dynamism (anticipated number of system changes over some time interval)
5. organizational culture (whether the organization thrives on chaos or order)

In Figure 2.8, as project characteristics tend away from the center of the diagram the likelihood of succeeding using agile methodologies decreases. Therefore, projects assessed in the innermost circle are likely candidates for agile approaches. Those in the second circle (but not in the inner circle) are marginal, and those outside of the second circle are not good candidates for agile approaches.

All of these process models look rather simplistic, artificial, or too prescriptive. Should they really be used?

It has been suggested by Parnas and Clements [1986] that after the project has been completed, tested, and delivered, users can "cover their tracks" by modifying the documentation so that it appears that a deliberate methodology was used. For example, when the sequence of the waterfall model cannot be followed strictly, at least the documentation should suggest that it was followed in that sequence.

While this kind of practice might appear disingenuous, the benefit is that a traceable history is established between each program feature and the requirement driving that feature. This approach promotes a maintainable, robust, and reliable product and, in particular, one where decisions related to timing requirements are well documented. It does indicate, however, that perhaps the process used was a reactive one and not part of a planned strategy.

2.4 Software Standards*

Who publishes software standards?

Standardizing organizations such as ISO, ACM, IEEE, the U.S. Department of Defense (DOD), and others actively promote the development, use, and improvement of standards for software processes and inherent life-cycle models. Even though many are interrelated and mutually influenced, the array of standards available can be confusing and even contradictory to the point of frustration.

What is the DOD-STD-2167A standard?

This extinct standard had a great deal of influence on the development of military software systems in the 1980s and 1990s. Because the U.S. DOD is the single largest procurer of software, the 2167A and waterfall "culture" pervades suppliers of military systems software even today, and so it is worth discussing briefly.

DOD-STD-2167A** was designed to produce documentation that achieves a high-integrity description of the evolving software design for baseline control and that serves as the foundation for life-cycle management. Formal reviews were prescribed throughout, but were sometimes just staged presentations. These audits often proved to be of questionable value and ultimately increased the cost of the system.

However, DOD-STD-2167A provided structure and discipline for the chaotic and complex development environment of large and mission-critical embedded applications.

What is the DOD-STD-498 standard?

This is another extinct standard, though vestiges of it too can be found widely throughout the defense and other industries. DOD-STD-498 was a merger of DOD-STD-2167A, used for weapon systems, with DOD-STD-7935A, used for automated information systems. Together, they formed a single software development standard for all of the organizations in the purview of the U.S. DOD. The purpose of developing this new standard, which was approved in 1994, was to resolve issues raised in the use of the

---

* Some of the following discussion is adapted from the excellent text on software standards by Wang and King [2000].

** DOD standards are sometimes referred to as "MIL-STD," for "military standard." So "DOD-STD-498" is equivalent to "MIL-STD-498."

old standards, particularly with their incompatibility with modern software engineering practice.

The process model adopted in DOD-STD-498 was significantly different from 2167A. The former standard explicitly imposed a waterfall model, whereas 498 provided for a development model that is compatible with all of the software life-cycle models discussed previously, except the lightweight methodologies.

What is the ISO 9000 standard?

ISO (International Organization for Standardization) 9000 is a generic, worldwide standard for quality improvement. The standard, which collectively is described in five standards (ISO 9000 through ISO 9004), was designed to be applied in a wide variety of manufacturing environments. ISO 9001 through ISO 9004 apply to enterprises according to the scope of their activities. ISO 9004 and ISO 9000-X family are documents that provide guidelines for specific applications domains.

Which part of ISO 9000 applies to software?

For software development, ISO 9000-3 is the document of interest. ISO released the 9000-3 quality guidelines in 1997 to help organizations apply the ISO 9001(1994) requirements to computer software. ISO 9000-3 is essentially an expanded version of ISO 9001 with added narrative to encompass software.

Who uses this standard?

ISO 9000-3 is widely adopted in Europe, and an increasing number of U.S. and Asian companies have adopted it as well.

What is in the ISO 9000-3 standard?

The ISO standards are process-oriented, "common-sense" practices that help companies create a quality environment. The principal areas of quality focus are:

Management responsibility
Quality system requirements
Contract review requirements
Product design requirements
Document and data control
Purchasing requirements
Customer supplied products
Product identification and tractability
Process control requirements
Inspection and testing
Control of inspection, measuring, and test equipment

Inspection and test status
Control of nonconforming products
Corrective and preventive actions
Handling, storage, and delivery
Control of quality records
Internal quality audit requirements
Training requirements
Servicing requirements
Statistical techniques

Paying particular attention to some of these areas, such as inspection and testing, design control, and product traceability (through a "rational design process"), can increase the quality of a software product.

How specific is ISO 9000-3 for software?

Unfortunately, the standard is very general and provides little specific process guidance. For example, Figure 2.9 illustrates ISO 9000-3: 4.4 Software development and design. While these recommendations are helpful as a "checklist," they provide very little in terms of a process that can be used.

While a number of metrics have been available to add some rigor to this somewhat generic standard, in order to achieve certification under the ISO standard, significant paper trails and overhead are required.

What is ISO/IEC standard 12207?

ISO 12207: Standard for Information Technology — Software Life Cycle Processes, describes five "primary processes"— acquisition, supply, devel-

ISO 9000-3	4.4 Software development and design
4.4.1 General	Develop and document procedures to control the product design and development process. These procedures must ensure that all requirements are being met.

Software development	Control your software development project and make sure that it is executed in a disciplined manner.
	• Use one or more life cycle models to help organize your software development project.
	• Develop and document your software development procedures. These procedures should ensure that:
	  • Software products meet all requirements.
	  • Software development follows your:
	    • Quality plan.
	    • Development plan.

FIGURE 2.9
Excerpt from ISO 9000-3: 4.4 Software development and design.

opment, maintenance, and operation. ISO 12207 divides the five processes into "activities," and the activities into "tasks," while placing requirements upon their execution. It also specifies eight "supporting processes" — documentation, configuration management, quality assurance, verification, validation, joint review, audit, and problem resolution — as well as four "organizational processes" — management, infrastructure, improvement, and training.

The ISO standard intends for organizations to tailor these processes to fit the scope of their particular projects by deleting all inapplicable activities, and it defines ISO 12207 compliance as the performance of those processes, activities, and tasks selected by tailoring.

ISO 12207 provides a structure of processes using mutually accepted terminology, rather than dictating a particular life-cycle model or software development method. Because it is a relatively high-level document, ISO 12207 does not specify the details of how to perform the activities and tasks comprising the processes. Nor does it prescribe the name, format, or content of documentation. Therefore, organizations seeking to apply ISO 12207 need to use additional standards or procedures that specify those details.

The IEEE recognizes this standard with the equivalent numbering "IEEE/EIA 12207.0-1996 IEEE/EIA Standard Industry Implementation of International Standard ISO/IEC12207:1995 and (ISO/IEC 12207) Standard for Information Technology—Software Life Cycle Processes."

2.5 Further Reading

Ambler, S., "Are you Agile or Fragile?" presentation, http://www.whysmalltalk.com/Smalltalk_Solutions/ss2003/pdf/ambler.pdf last accessed April 10, 2006.

Beck, K., Embracing change with extreme programming, *Computer*, 32(10), 70–77, 1999.

Boehm, B.W., A spiral model of software development and enhancement, *Computer*, 21(5), 61–72, 1988.

Boehm, B.W. and Turner, R., *Balancing Agility and Discipline: A Guide to the Perplexed*, Addison-Wesley, Boston, MA, 2004.

Brooks, F.P., *The Mythical Man-Month*, 20th Anniversary Edition, Addison-Wesley, Boston, MA, 1995.

Institute of Electrical and Electronic Engineers (IEEE), *IEEE 1473-1999, IEEE Standard for Communications Protocol Aboard Trains*, 1999.

Laplante, P.A., *Software Engineering for Image Processing Systems*, CRC Press, Boca Raton, FL, 2004.

Martin, R.C., *Agile Software Development: Principles, Patterns, and Practices*, Prentice-Hall, Englewood Cliffs, NJ, 2002.

Neill, C.J. and Laplante, P.A., Requirements engineering: the state of the practice, *Software*, 20(6), 40–46, 2003.

Nerur, S., Mahapatra, R., and Mangalaraj, G., Challenges of migrating to agile methodologies, *Commun. ACM*, 48(5), 73–78, 2005.

Nord , R.L. and Tomayko, J.E., Software architecture-centric methods and agile development, *Software*, 23(2), 47–53, 2006.

Parnas, D.L. and Clements, P. C., A rational design process: how and why to fake it, *IEEE Trans. Software Eng.*, 12(2), 251–257, 1986.

Reifer, D., How good are agile methods, *Software*, 19(4), 16–18, 2002.

Royce, W., *Software Project Management: A Unified Framework*, Addison-Wesley, Boston, MA, 1998.

Theuerkorn, F., *Lightweight Enterprise Architectures*, Auerbach Publications, Boca Raton, FL, 2005.

Tucker, A.B., Jr. (Editor-in-Chief), *The Computer Science and Engineering Handbook*, CRC Press, Boca Raton, FL, 1996.

Wang, Y.W., and King, G., *Software Engineering Processes: Principles and Applications*, CRC Press, Boca Raton, FL, 2000.

"The Rules and Practices of Extreme Programming," www.extremeprogramming.org/rules.html, accessed September 14, 2006.

3

Software Requirements Specification

Outline

- Requirements engineering concepts
- Requirements specifications
- Requirements elicitation
- Requirements modeling
- Requirements documentation
- Recommendations on requirements

3.1 Introduction

Requirements engineering is the process of eliciting, documenting, analyzing, validating, and managing requirements. Different approaches to requirements engineering exist, some more complete than others. Whatever the approach taken, it is crucial that there is a well-defined methodology and that documentation exists for each stage.

Requirements modeling involves the techniques needed to express requirements in a way that can capture user needs. Requirements modeling uses techniques that can range from high-level abstract statements through psuedocode-like specifications, formal logics, and graphical representations. Whatever representation technique is used, the requirements engineer must always strive to gather complete, precise, and detailed specifications of system requirements.

This chapter incorporates a discussion of these aspects of requirements engineering. The chapter concludes with a discussion on requirements documentation and with some recommendations and best practices.

3.2 Requirements Engineering Concepts

What is software requirements engineering?

Requirements engineering is a subdiscipline of software engineering that is concerned with determining the goals, functions, and constraints of software systems. Requirements engineering also involves the relationship of these factors to precise specifications of software behavior, and to their evolution over time and across software families.

When does requirements engineering start?

Ideally, the requirements engineering process begins with a feasibility study activity, which leads to a feasibility report. It is possible that the feasibility study may lead to a decision not to continue with the development of the software product. If the feasibility study suggests that the product should be developed, then requirements analysis can begin.

What is software requirements specification?

This is the set of activities designed to capture behavioral and nonbehavioral aspects of the system in the SRS document. The goal of the SRS activity, and the resultant documentation, is to provide a complete description of the system's behavior without describing the internal structure. This aspect is easily stated, but difficult to achieve, particularly in those systems where temporal behavior must be described.

Why do we need SRSs?

Precise software specifications provide the basis for analyzing the requirements, validating that they are the stakeholder's intentions, defining what the designers have to build, and verifying that they have done so correctly.

How do software requirements help software engineers?

SRSs allow us to know the motivation for development of the software system.

Software requirements also help software engineers manage the evolution of the software over time and across families of related software products. This approach reflects the reality of a changing world and the need to reuse partial specifications.

What are the core requirements engineering activities?

First, we must elicit the requirements, which is a form of discovery. Some people use the term "gathering" to describe the process of collecting software requirements, but "gathering" implies that requirements are like vegetables in the garden to be harvested. Often, though, requirements are deeply hidden, expressed incorrectly by stakeholders, contradictory, and complex. Eliciting requirements is one of the hardest jobs for the requirements engineer.

Modeling requirements involves representing the requirements in some form. Words, pictures, and mathematical formulas can all be used but it is never easy to effectively model requirements.

Analyzing requirements involves determining if the requirements are correct or have certain other properties such as consistency, completeness, sufficient detail, and so on. The requirements models and their properties must also be communicated to stakeholders and the differences reconciled.

Finally, requirements change all the time and the requirements engineer must be equipped to deal with this eventuality. We will focus our discussions on these aforementioned areas.

Which disciplines does requirements engineering draw upon?

Requirements engineering is strongly influenced by computer science and systems engineering. But because software engineering is a human endeavor, particularly with respect to understanding people's needs, requirements engineering draws upon such diverse disciplines as philosophy, cognitive psychology, anthropology, sociology, and linguistics.

What is a requirement?

A requirement can range from a high-level, abstract statement of a service or constraint to a detailed, formal specification. There is so much variability in requirements detail because of the many purposes that the requirements must serve.

3.3 Requirements Specifications

What kinds of SRSs are there?

SRSs are usually classified in terms of their level of abstraction:

- user requirements
- system requirements
- software design specifications [Sommerville 2005].

What are user requirements specifications?

User requirements specifications are usually used to attract bidders, often in the form of a request for proposal (RFP).

User requirements may contain abstract statements in natural language, for example, English, with accompanying informal diagrams, even back-of-the-napkin drawings*. User requirements specify functional and nonfunctional requirements as they pertain to externally visible behavior in a form understandable by clients and system users. These kinds of representation techniques, however, are fraught with danger, as will be discussed shortly.

What are system requirements specifications?

System level requirements, or SRSs mentioned previously, are detailed descriptions of the services and constraints. Systems requirements are derived from analysis of the user requirements. Systems requirements should be structured and precise because they act as a contract between client and contractor (and can literally be enforced in court). Appendix A contains an SRS for a wet well control system. This specification is nearly complete (it is lacking some elements, mostly formal specifications and some narrative that are omitted for brevity). I will be referring to this document occasionally, and now would be a good time for you to browse the first few pages of Appendix A to become familiar with this simple engineering application.

What are software design specifications?

Software design specifications are usually the most detailed level requirements specifications that are written and used by software engineers as a basis for the system's architecture and design. Appendix B contains the corresponding software design specification for the wet well control system, which was derived from Appendix A. The software design specification is described in Chapter 4.

Within these three specification types are there different requirements types?

Yes, there are. Within the family of user, system, and functional requirements specifications, all kinds of things can be described. These include external constraints to the software and user needs. The user needs are usually called "functional requirements" and the external constraints are called "nonfunctional requirements."

What are functional requirements?

Functional requirements describe the services the system should provide. Sometimes the functional requirements state what the system should not do. Functional requirements can be high-level and general or detailed, expressing inputs, outputs, exceptions, and so on. See Appendix A, Section A.3 for specific functional requirements for the wet well control system.

* The Compaq portable computer, one of the first of its kind, was conceptualized on a paper restaurant placemat.

What are nonfunctional requirements?

Nonfunctional requirements are imposed by the environment in which the system is to exist. These requirements could include timing constraints, quality properties, standard adherence, programming languages to be used, compliance with laws, and so on.

What are domain requirements?

Domain requirements are a type of nonfunctional requirement from which the application domain dictates or derives. Domain requirements might impose new functional requirements or constraints on existing functional requirements.

For example, in the baggage inspection system, industry standards and restrictions on baggage size and shape will place certain constraints on the system.

What are interface specifications?

Interface specifications are functional software requirements specified in terms of interfaces to operating units. Most systems must operate with other systems and the operating interfaces must be specified as part of the requirements. There are three types of interface that may have to be defined:

1. procedural interfaces
2. data structures that are exchanged
3. data representations

Formal notations are an effective technique for interface specification

What are performance requirements?

Performance requirements are functional requirements that include the static and dynamic numerical requirements placed on the software or on human interaction with the software as a whole. For an imaging system, static requirements might include the number of simultaneous users to be supported. The dynamic requirements might include the numbers of transactions and tasks the amount of data to be processed within certain time periods for both normal and peak workload conditions. For the wet well control system in Section A.3.2, many of the requirements must be achieved within five seconds. These time constraints represent performance requirements.

What are logical database requirements?

Logical database requirements are functional requirements that include the types of information used by various functions such as frequency of use, accessing capabilities, data entities and their relationships, integrity constraints, and data retention requirements.

What are design constraint requirements?

Design constraint requirements are nonfunctional requirements that are related to standards compliance and hardware limitations.

What are system attribute requirements?

System attribute requirements are functional requirements that include reliability, availability, security, maintainability, and portability. Many of these requirements can be found in Appendix A.

What is a feasibility study and what is its role?

A feasibility study is a short focused study that checks if the system contributes to organizational objectives the system can be engineered using current technology and within budgetthe system can be integrated with other systems that are used.

A feasibility study is used to decide if the proposed system is worthwhile. Feasibility studies can also help answer some of the following questions.

- What if the system wasn't implemented?
- What are current process problems?
- How will the proposed system help?
- What will be the integration problems?
- Is new technology needed? What skills?
- What facilities must be supported by the proposed system?

Therefore, it is important to conduct a feasibility study before a large investment is made in a system that is not needed or which does not solve some inherent business process problem.

Are their social and organizational factors in requirements engineering?

You bet there are. Software engineering involves many human activities, and none is more obviously so than requirements engineering. Software systems are used in a social and organizational context; thus, political factors almost always influence the software requirements.

Moreover, stakeholders often don't know what they really want and they express requirements in their own terms. Different stakeholders may have conflicting requirements, new stakeholders may emerge, and the business environment may change. Finally, the requirements change during the analysis process for a host of reasons. Therefore, good requirements engineers must be sensitive to these factors.

3.4 Requirements Elicitation

What is requirements elicitation?

Requirements elicitation involves working with customers to determine the application domain, the services that the system should provide, and the operational constraints of the system. Elicitation may involve end-users, managers, engineers involved in maintenance, domain experts, trade unions, and so on. These people are collectively called stakeholders.

But stakeholders don't always know what they want, right?

True. An important consideration in eliciting requirements from stakeholders is that they often don't know what they really want. The software engineer has to be sensitive to the needs of the stakeholders and aware of the problems that stakeholders can create including:

- expressing requirements in their own terms
- providing conflicting requirements
- introducing organizational and political factors, which may influence the system requirements
- changing requirements during the analysis process due to new stakeholders who may emerge and changes to the business environment

The software engineer must monitor and control these factors throughout the requirements engineering process.

Are there any practical approaches to requirements elicitation?

Yes, there are several. The following three approaches will be discussed in detail:

- Joint application design (JAD)
- Quality function deployment (QFD)
- Designer as apprentice

Each of these techniques has been used successfully for requirements elicitation.

What is JAD?

JAD involves highly structured group meetings or mini-retreats with system users, system owners, and analysts in a single room for an extended period. These meetings occur four to eight hours per day and over a period lasting one day to a couple of weeks.

JAD and JAD-like techniques are becoming increasingly common in systems planning and systems analysis to obtain group consensus on problems, objectives, and requirements. Specifically, software engineers can use JAD for:

- eliciting requirements and for the SRS
- design and software design description
- code
- tests and test plans
- user manuals

There can be multiple reviews for each of these artifacts, if necessary.

How do you plan for a JAD session?

Planning for a review or audit session involves three steps:

selecting participants
preparing the agenda
selecting a location

Reviews and audits may include some or all of the following participants:

- sponsors (for example, senior management)
- a team leader (facilitator, independent)
- users and managers who have ownership of requirements and business rules
- scribes
- engineering staff

The sponsor, analysts, and managers select a leader. The leader may be inhouse or a consultant. One or more scribes (note-takers) are selected, normally from the software development team. The analyst and managers must select individuals from the user community. These individuals should be knowledgeable and articulate in their business area.

Before planning a session, the analyst and sponsor determine the scope of the project and set the high-level requirements and expectations of each session. The session leader also ensures that the sponsor is willing to commit people, time, and other resources to the effort. The agenda depends greatly on the type of review to be constructed and should be created to allow for sufficient time. The agenda, code, and documentation is then sent to all participants well in advance of the meeting so that they have sufficient time to review them, make comments, and prepare questions.

What are some of the ground rules for JAD sessions?

The following are some rules for conducting software requirements, design audits, or code walkthrough. The session leader must make every effort to ensure that these practices are implemented.

- Stick to the agenda.
- Stay on schedule (agenda topics are allotted specific time).
- Ensure that the scribe is able to take notes.
- Avoid technical jargon (if the review is a requirements review and involves nontechnical personal).
- Resolve conflicts (try not to defer them).
- Encourage group consensus.

- Encourage user and management participation without allowing individuals to dominate the session.
- Keep the meeting impersonal.

The end product of any review session is typically a formal written document providing a summary of the items (specifications, design changes, code changes, and action items) agreed upon during the session. The content and organization of the document obviously depends on the nature and objectives of the session. In the case of requirements elicitation, however, the main artifact could be a first draft of the SRS.

What is QFD?

QFD was introduced by Yoji Akao in 1966 for use in manufacturing, heavy industry, and systems engineering. It is a technique for determining customer requirements and defining major quality assurance points to be used throughout the production phase. QFD provides a structure for ensuring that customers' wants and needs are carefully heard, and then directly translated into a company's internal technical requirements — from analysis through implementation to deployment. The basic idea of QFD is to construct relationship matrices between customer needs, technical requirements, priorities, and (if needed) competitor assessment. Because these relationship matrices are often represented as the roof, ceiling, and sides of a house, QFD is sometimes referred to as the "house of quality" (Figure 3.1) [Akao 1990]. It has been applied to software systems by IBM, DEC, HP, AT&T, Texas Instruments and others.

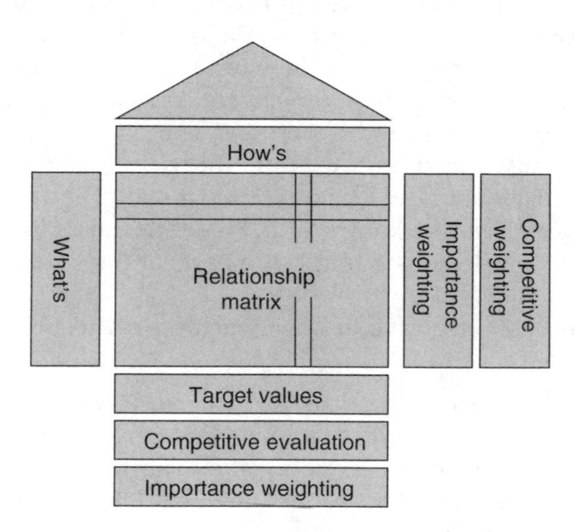

FIGURE 3.1
QFD's "house of quality." (From Akao, Y. (Ed.), *Quality Function Deployment*, Productivity Press, Cambridge, MA, 1990.)

What is the "voice of the customer?"

This is what customers want and need from the product, stated in their words as much as possible. The voice of the customer forms the basis for all analysis, design, and development activities to ensure that products are not developed from "the voice of the engineer" only. This approach embodies the essence of requirements elicitation.

What are the advantages of using QFD?

QFD improves the involvement of users and mangers. It shortens the development life cycle and improves overall project development. QFD supports team involvement by structuring communication processes. Finally, it provides a preventive tool to avoid the loss of information.

Are there drawbacks to using QFD for software requirements discovery?

Yes. For example, there may be difficulties in expressing temporal requirements. QFD is difficult to use with an entirely new project. For example, how do you discover customer requirements for something that does not exist and how do you build and analyze the competitive products?

Sometimes it is hard to find measurements for certain functions and to keep the level of abstraction uniform. And, the less we know the less we document. Finally, sometimes the house of quality can become a damn big mansion; that is, the desirable features list can get out of control.

What is "designer as apprentice?"

Designer as apprentice is a requirements discovery technique in which the requirements engineer "looks over the shoulder" of the customer to enable the engineer to learn enough about the work of the customer to understand his needs. The relationship between customer and designer is like that between a master craftsman and apprentice. That is, the apprentice learns a skill from the master just as we want the requirements engineer (the designer) to learn about the work from the customer. The apprentice is there to learn whatever the master knows (and, therefore, must guide the customer in talking about and demonstrating those parts of the work).

But doesn't the customer have to have teaching ability for this technique to work?

No. Some customers cannot talk about their work effectively, but can talk about it as it unfolds. Moreover, customers don't have to work out the best way to present it, or the motives; they just explain what they are doing.

Seeing the work reveals what matters. For example, people are not aware of everything they do and sometimes why they do it. Some actions are the result of years of experience and are too subtle to express. Other actions are just habits with no valid justification. The presence of an apprentice provides the opportunity for the master (customer) to think about the activities and how they come about.

Seeing the work reveals details; unless we are performing a task, it is difficult to describe it in detail. Finally, seeing the work reveals structure. Patterns of working are not always obvious to the worker. An apprentice learns the strategies and techniques of work by observing multiple instances of a task and forming an understanding of how to do it himself by incorporating the variations.

Who is responsible for seeing the work structure in this technique?

The designer must understand the structure and implication of the work, including:

- the strategy to get work done
- constraints that get in the way
- the structure of the physical environment as it supports work
- the way work is divided
- recurring patterns of activity
- the implications the above has on any potential system

The designer must demonstrate to the customer his understanding of the work so that any misunderstandings can be corrected.

Does design as apprentice have any other benefits?

Yes. In fact, using this technique can help improve the modeled process. Both customer and designer learn during this process; the customer learns what may be possible and the designer expands his understanding of the work. If the designer has an idea for improving the process, however, this must be fed back to the customer immediately.

3.5 Requirements Modeling

How are software requirements modeled?

There are a number of ways to model software requirements; these include natural languages, informal and semiformal techniques, user stories, use case diagrams, structured diagrams, object-oriented techniques, formal methods, and more. We will discuss some of these in detail.

Why can't requirements just be communicated in English?

English, or any other natural language, is fraught with problems for requirements communication. These problems include lack of clarity and precision, mixing of functional and nonfunctional requirements, and requirements amalgamation, where several different requirements may be expressed together. Other problems with natural languages include ambiguity, over-flexibility, and lack of modularization.

These shortcomings, however, do not mean that natural language is never used in an SRS. Every clear SRS must have a great deal of narrative in clear and concise natural language. But when it comes to expressing complex behavior, it is best to use formal or semiformal methods, clear diagrams or tables, and narrative as needed to tie these elements together.

What are some alternatives to using natural languages?

Alternatives to natural languages include:

- structured natural language
- design description languages
- graphical notations
- mathematical specifications

What is a structured language specification?

This is a limited form of natural language that can be used to express requirements. In other words, the vocabulary and grammar rules available to express requirements are strictly controlled and are capable of being parsed.

Structured languages remove some of the problems resulting from ambiguity and flexibility and impose a degree of uniformity on a specification. On the other hand, use of structured languages requires a level of training that can frustrate stakeholders. To increase usability, structured languages can be facilitated using a forms-based approach.

In a forms-based approach, templates or forms are created and given to the customer and other stakeholders to be filled in. The template can include the following information:

definition of the function or entity

description of inputs and where they come from

description of outputs and where they go to

indication of other entities required

pre- and postconditions (if appropriate)

side effects (if any)

The data in the template can then be converted to structured language using an appropriate interpreter. Figure 3.2 depicts a simple example of a structured language template. Forms-based specification can still be used even when structured languages are not being used.

Function	
Description	
Inputs	
Outputs	
Destination	
Requires	
Precondition	
Postcondition	
Side-effects	

FIGURE 3.2
A forms-based specification template.

What are program design language-based requirements?

Program design languages (PDLs) involve requirements that are defined operationally using a language like a programming language but with more flexibility of expression. PDLs are most appropriate in two situations:

1. where an operation is specified as a sequence of actions and the order is important
2. when hardware and software interfaces have to be specified

Are there any disadvantages to using PDLs?

There are a few. For example, the PDL may not be sufficiently expressive to define domain concepts. Moreover, the specification may be taken as a design rather than a specification. Finally, the notation may be understandable only to people with knowledge of programming languages.

What are use cases?

Use cases are an essential artifact in object-oriented requirements elicitation and analysis and are described graphically using any of several techniques. One representation for the use case is the use case diagram, which depicts the

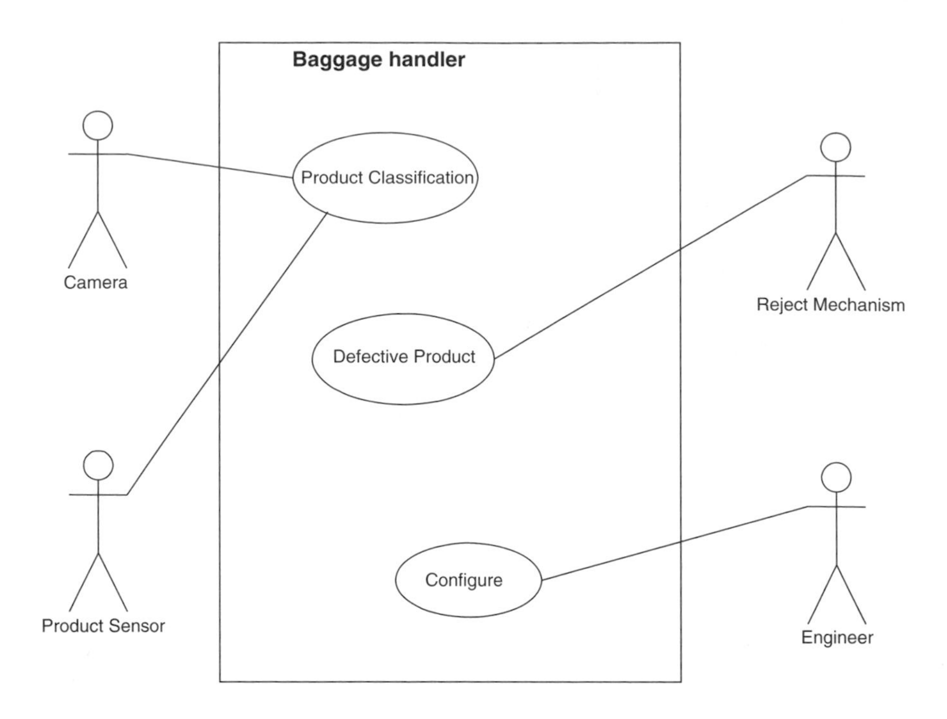

FIGURE 3.3
Use case diagram of the baggage inspection system.

interactions of the software system with its external environment. In a use case diagram, the box represents the system itself. The stick figures represent "actors" that designate external entities that interact with the system. The actors can be humans, other systems, or device inputs. Internal ellipses represent each activity of use for each of the actors (use cases). The solid lines associate actors with each use. Figure 3.3 shows a use case diagram for the baggage inspection system.

Figure A.4 in Appendix A shows a use case diagram for the wet well control system.

Each use case is, however, a document that describes scenarios of operation of the system under consideration as well as pre- and postconditions and exceptions. In an iterative development life cycle, these use cases will become increasingly refined and detailed as the analysis and design workflows progress. Interaction diagrams are then created to describe the behaviors defined by each use case. In the first iteration these diagrams depict the system as a "black box," but once domain modeling has been completed the black box is transformed into a collaboration of objects as will be seen later.

What are user stories?

User stories are short conversational texts that are used for initial requirements discovery and project planning. User stories are widely used in conjunction with agile methodologies.

User stories are written by the customers in their own "voice," in terms of what the system needs to do for them. User stories usually consist of two to four sentences written by the customer in his own terminology, usually on a three-by-five inch card. The appropriate amount of user stories for one system increment or evolution is about 80, but the appropriate number will vary widely depending upon the application size and scope.

An example of a user story for the wet well system described in Appendix A through Appendix C is as follows:

> There are two pumps in the wet well control system. The control system should start the pumps to prevent the well from overflowing. The control system should stop the pumps before the well runs dry.

User stories should provide only enough detail to make a reasonably low-risk estimate of how long the story will take to implement. When the time comes to implement the story, developers will meet with the customer to flesh out the details.

User stories also form the basis of acceptance testing. For example, one or more automated acceptance tests can be created to verify the user story has been correctly implemented.

What are formal methods in software specification?

Formal methods attempt to improve requirements formulation and expression by applying mathematics and logic. Formal methods employ some combination of predicate calculus (first order logic), recursive function theory, Lambda calculus, programming language semantics, discrete mathematics, number theory, and abstract algebra. This approach is attractive because it offers a more scientific method for requirements specification.

By their nature, specifications for most embedded systems usually contain some formality in the mathematical expression of the underlying imaging operations.

What are the motivations for using formal methods?

One of the primary attractions of formal methods is that they offer a highly scientific approach to development. Formal requirements offer the possibility of discovering errors at the earliest phase of development, while the errors can be corrected quickly and at a low cost. Informal specifications might not achieve this goal because they are not precise enough to be refuted by finding counter examples.

What are informal and semiformal methods?

Approaches to requirements specification that are not formal are either informal (such as flow-charting) or semiformal. The UML is a semiformal specification approach, meaning that while it does not appear to be mathematically based, it is in fact nearly formal in that every one of its modeling tools can be converted either completely or partially to an underlying mathematical representation (a work group is focused on remedying these deficiencies). In any case, UML largely enjoys the benefits of both informal and formal techniques.

How are formal methods used?

Formal methods are typically not intended to take on an all-encompassing role in system or software development. Instead, individual techniques are designed to optimize one or two parts of the development life cycle.

There are three general uses for formal methods:

1. Consistency checking — system behavioral requirements are described using a mathematically based notation.
2. Model checking — state machines are used to verify if a given property is satisfied under all conditions.
3. Theorem proving — axioms of system behavior are used to derive a proof that a system will behave in a given way.

Formal methods offer important opportunities for reusing requirements. Embedded systems are often developed as families of similar products, or as incremental redesigns of existing products. For the first situation, formal methods can help identify a consistent set of core requirements and abstractions to reduce duplicate engineering effort. For redesigns, having formal specifications for the existing system provides a precise reference for baseline behavior, and provides a way to analyze proposed changes.

Are formal methods hard to use?

Formal methods can be difficult to use and are sometimes error-prone. For these reasons and because they are sometimes perceived to increase early life-cycle costs and delay projects, formal methods are frequently and unfortunately avoided.

What are some of the formal methods techniques?

Formal methods include Z, Vienna design method (VDM), and communicating sequential processes (CSP). All of these methods are highly specialized and require a great deal of formal mathematical training that most traditional engineers do not receive.

I will briefly introduce one of the most celebrated formal methods, Z. But I am really going to focus on some more familiar mathematical models that, while not usually considered as high-flying as these other formal methods, are formal methods just the same in that they are mathematically based. These methods also have the advantages of being known to most engineers and can be used in both software specification and design.

What is Z?

Z (pronounced "zed"), introduced in 1982, is a formal specification language that is based on set theory and predicate calculus. As in other algebraic approaches, the final specification in Z is reached by a refinement process starting from the most abstract aspects of the systems. There is a mechanism for system decomposition known as the Schema Calculus. Using this calculus, the system specification is decomposed in smaller pieces called schemes where both static and dynamic aspects of system behavior are described.

The Z language does not have any support for defining timing constraints. Therefore, in recent years, several extensions for time management have been proposed. For example, Z has been integrated with real-time interval logic (RTIL), which provides for an algebraic representation of temporal behavior.

There are other extensions of Z to accommodate the object-oriented approaches, which adds formalism for modularity and specification reuse. These extensions define the system state space as a composition of the state spaces of the individual system objects.

Most of these extensions also provide for information hiding, inheritance, polymorphism, and instantiation into the Z Schema Calculus.

For example, one extension, Object-Z, includes all the aforementioned extensions and further integrates the concepts of temporal logic, making it suitable for real-time specification. In this language the object status is an event history of object behavior making the language more operational than the early version of Z.

A Z-like formal description of the wet well control system can be found in Section 3.2.5 of Appendix A. Note that this is not a Z specification, as it is lacking much of the front-end preamble and declarations. However, the mathematical notation is very Z-like.

What are finite state machines?

The finite state automaton (FSA), finite state machine (FSM), or state transition diagram (STD) are types of mathematical models used in the specification and design of a wide range of systems. Intuitively, FSMs rely on the fact that a fixed number of unique states can represent many systems. The system may change state depending upon time or the occurrence of specific events — a fact that is reflected in the automaton.

How are FSAs represented?

An FSM can be specified in diagrammatic, set-theoretic, and matrix representations. To illustrate them, consider the baggage inspection system. Suppose it can be in one of three modes of operation: calibration, diagnostic, or operational. The calibration mode is entered when the operator sets a signal (op_cal). Similarly, the system returns to operational mode upon issuance of the op_op signal. The diagnostic mode is entered if an exceptional condition

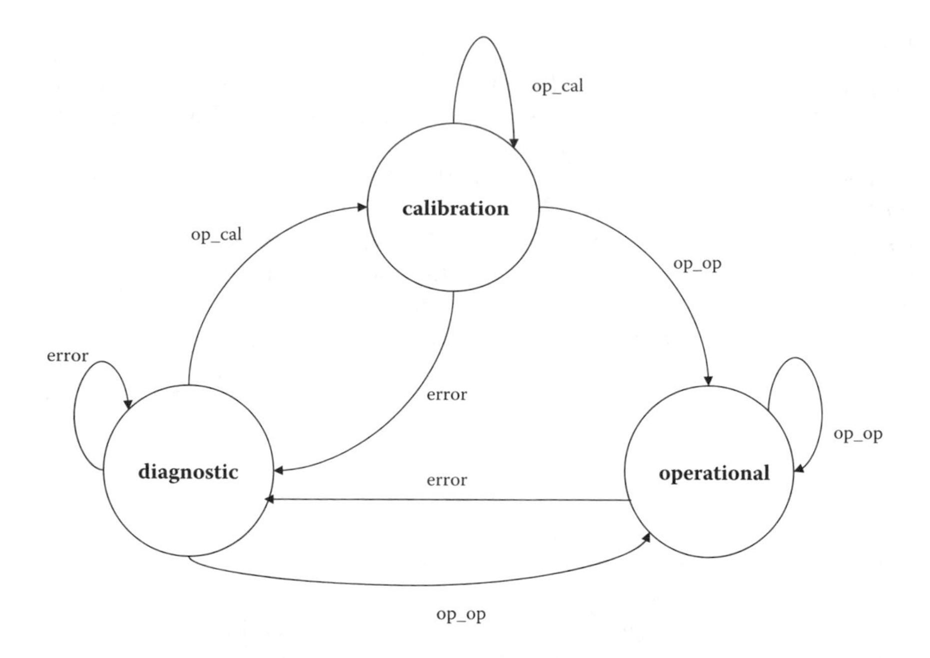

FIGURE 3.4
Partial FSM showing behavior of the baggage inspection system.

or error occurs in either of the other modes. The diagnostic mode can only be exited by operator intervention by setting the appropriate signal. This behavior can be described by the FSM shown in Figure 3.4.

The behavior shown in Figure 3.4 can also be represented mathematically by the five-tuple

$$M = \{S, i, T, \Sigma, \delta\} \quad (3.1)$$

where S is a finite, nonempty set of states, i is the initial state (i is a member of S), T is the set of terminal states (T is a subset of S), Σ is an alphabet of symbols or events used to mark transitions, δ is a transition function that describes the next state of the machine given the current state and a symbol from the alphabet (an event). That is, $\delta : S \times \Lambda \rightarrow S$

Can you give an example of an FSM?

In the baggage inspection system example, S = {calibration, diagnostic, operational}, i = calibration, $T = S$, and Σ = {op_op, op_cal, error}. The transition function can be described by a set of triples of the form (state, signal, next_state).

It is usually more convenient to represent the transition function with a transition table, as shown in Table 3.1.

TABLE 3.1

Transition Table for the FSM of the Baggage Inspection System Shown in Figure 3.4

		Event	
Current state	**op_op**	**op_cal**	**error**
Calibration	Operational	Calibration	Diagnostic
Diagnostic	Operational	Calibration	Diagnostic
Operational	Operational	Calibration	Diagnostic

Note: The internal entries represent the functional mode to be performed.

What is a Mealy FSM?

An FSM that does not depict outputs during transition is called a Moore machine. Outputs during transition can be depicted, however, by a variation of the Moore machine called a Mealy machine. The Mealy machine can be described mathematically by a six-tuple,

$$M = \{S, i, T, \Sigma, \Gamma, \delta\} \tag{3.2}$$

where the first five elements of the six-tuple are the same as for the Moore machine and a sixth parameter, Γ , which represents the set of outputs. The transition function is slightly different than previously in that it describes the next state of the machine given the current state, and a symbol from the alphabet. The transition function is then $\delta : S \times \Lambda \to S \times \Gamma$.

Can you give an example of a Mealy machine?

A general Mealy machine for a system with three states, three inputs, and three outputs such as the baggage inspection system is shown in Figure 3.5.

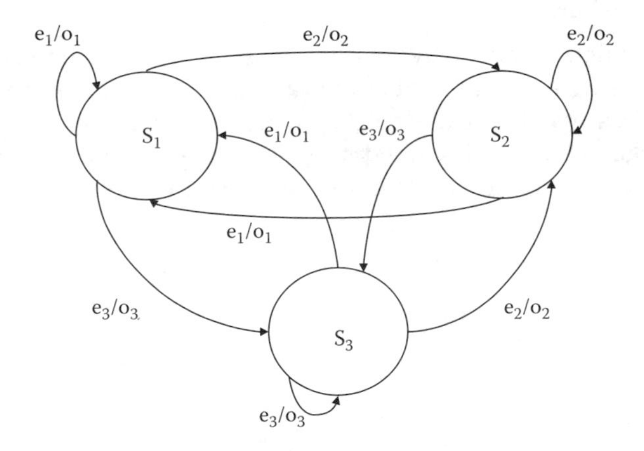

FIGURE 3.5
A generic Mealy machine for a three-state system with events e_1, e_2, e_3 and outputs o_1, o_2, o_3.

TABLE 3.2

Transition Matrix for the FSM in Figure 3.5

	S_1	S_2	S_3
e_1	S_1/S_1	S_1/S_1	S_1/S_1
e_2	S_2/O_2	S_2/O_2	S_2/O_2
e_3	S_3/O_3	S_3/O_3	S_3/O_3

The transition matrix for the FSM shown in Figure 3.5 is shown in Table 3.2.

What are the advantages of using FSMs in requirements specification or design?

FSMs are easy to visualize and to convert to a design and code. They are also unambiguous because they can be represented with a formal mathematical description. In addition, concurrency can be depicted by using multiple machines.

Finally, because mathematical techniques for reducing the number of states exist, programs based on FSMs can be formally optimized. A rich theory surrounds FSMs, and this can be exploited in the development of system specifications.

Are there any disadvantages to using FSMs?

The major disadvantage of FSMs is that the internal aspects, or "insideness" of modules cannot be depicted. That is, there is no way to indicate how functions can be broken down into subfunctions. In addition, intertask communication for multiple FSMs is difficult to depict. Finally, depending upon the system and alphabet used, the number of states can grow very large. However, these problems can be overcome using statecharts.

What are statecharts?

Statecharts, which are one of the base modeling languages in the UML family, combine FSMs with dataflow diagrams and a feature called broadcast communication in a way that can depict synchronous and asynchronous operations. Statecharts can be described succinctly as statecharts = FSM + depth + orthogonality + broadcast communication (Figure 3.6).

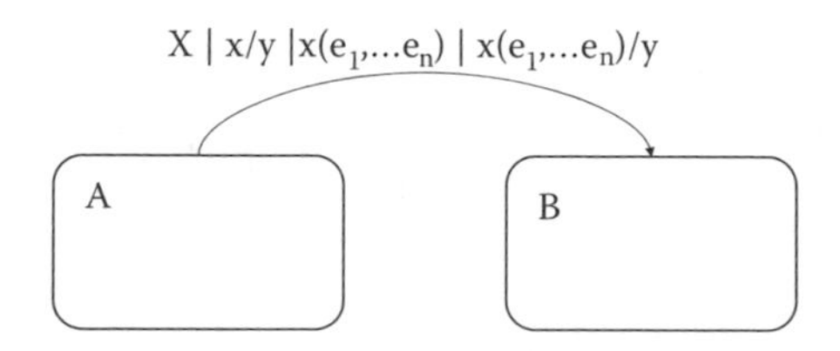

FIGURE 3.6
Statechart format where A and B are states, x is an event that causes the transition marked by the arrow, y is an optional event triggered by x, and $e_1,\ldots e_n$ are conditions qualifying the event.

Here, the FSM is a finite state machine, depth represents levels of detail, orthogonality represents separate tasks, and broadcast communication is a method for allowing different orthogonal processes to react to the same event. The statechart resembles an FSM where each state may contain its own FSM that describes its behavior. The various components of the statechart are depicted as follows:

The FSM is represented in the usual way, with capital letters or descriptive phrases used to label the states.

Depth is represented by the insideness of states.

Broadcast communications are represented by labeled arrows, in the same way as FSMs.

Orthogonality is represented by dashed lines separating states.

Symbols *a*, *b*, …, *z* represent events that trigger transitions, in the same way that transitions are represented in FSMs.

Small letters within parentheses represent conditions that must be true for the transitions to occur.

What is the advantage of using statecharts over FSMs?

A significant feature of statecharts is the encouragement of top-down design of a module. For example, for any module (represented like a state in an FSM), increasing detail is depicted as states internal to it. In Figure 3.7, the system is composed of states A and B. Each of these, in turn, can be decomposed into states A_1, A_2 and B_1, B_2, respectively, which might represent program modules.

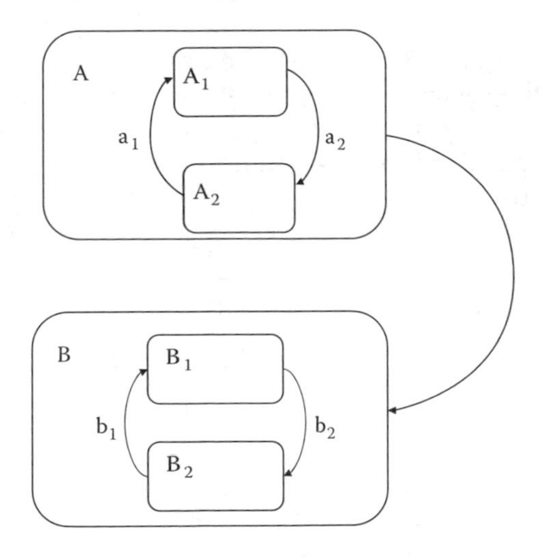

FIGURE 3.7
A statechart depicting insideness.

Those states can also be decomposed, and so forth. To the software designer, each nested substate within a state represents a procedure within a procedure.

What is orthogonality?

Orthogonality depicts concurrency in the system for processes that run in isolation, called AND states. Orthogonality is represented by dividing the orthogonal components by dashed lines. For example, if state Y consists of AND components A and D, Y is called the orthogonal product of A and D. If Y is entered from the outside (without any additional information), then the states A and D are entered simultaneously. Communication between the AND states can be achieved through global memory, whereas synchronization can be achieved through a unique feature of statecharts called broadcast communication.

What is broadcast communication?

Broadcast communication is depicted by the transition of orthogonal states based on the same event. For example, if an imaging system switches from standby to ready mode, an event indicated by an interrupt can cause a state change in several processes.

What is a chain reaction?

A unique aspect of broadcast communication is that it can signal a chain reaction; that is, events can trigger other events. The implementation follows from the fact that statecharts can be viewed as an extension of Mealy machines, and output events can be attached to the triggering event. In contrast with Mealy machines, however, the output is not seen by the outside world; instead, it affects the behavior of an orthogonal component.

For example, in Figure 3.8 suppose there exists a transition labeled *e*/*f* and if event *e* occurs, then event *f* is immediately activated. Event *f* could, in turn, trigger a transaction such as *f/g*. The length of a chain reaction is the number of transitions triggered by the first event. Chain reactions are assumed to occur instantaneously.

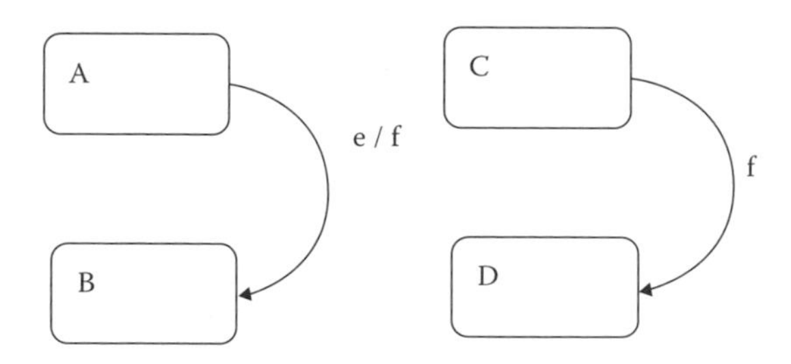

FIGURE 3.8
Statechart depicting a chain reaction.

In this system, a chain reaction of length two will occur when the *e/f* transition occurs.

When are statecharts useful in capturing requirements?

Statecharts are excellent for representing embedded systems because they can easily depict concurrency while preserving modularity. In addition, the concept of broadcast communication allows for easy intertasking.

In short, the statechart combines the best of dataflow diagrams and FSMs. Commercial products allow an engineer to graphically design a real-time system using statecharts, perform detailed simulation analysis, and generate Ada or C code. Furthermore, statecharts can be used in conjunction with both structured analysis (SA) and object-oriented analysis (OOA).

What are petri nets?

Petri nets are another formal method used to specify the operations to be performed in a multiprocessing or multitasking environment. While they have a rigorous foundation, they can also be described graphically. A series of circular bubbles called "places" are used to represent data stores or processes. Rectangular boxes are used to represent transitions or operations. The processes and transitions are labeled with a data count and transition function, respectively, and are connected by unidirectional arcs.

The initial graph is labeled with markings given by m_o, which represent the initial data count in the process. Net markings are the result of the firing of transitions. A transition, *t*, fires if it has as many inputs as required for output.

In petri nets, the graph topology does not change over time; only the "markings" or contents of the places do. The system advances as transitions "fire."

Can you give an example?

To illustrate the notion of "firing," consider the petri nets given in Figure 3.9 with the associated firing table given in Table 3.3.

As a somewhat more significant example, consider the petri net in Figure 3.10. Reading from left to right and top to bottom indicates the stages of firings in the net.

Table 3.4 depicts the firing table for the petri net in Figure 3.10.

Petri nets can be used to model systems and to analyze timing constraints and race conditions. Certain petri net subnetworks can model familiar flowchart constructs. Figure 3.11 illustrates these analogies.

How do I relate a petri net to software program behavior?

One way is to look at the low-level analogies of petri net configurations to familiar flowchart behavior as shown in Figure 3.11.

When are petri nets used in requirements analysis and specification?

Petri nets are excellent for representing multiprocessing and multiprogramming systems, especially where the functions are simple. Because they are

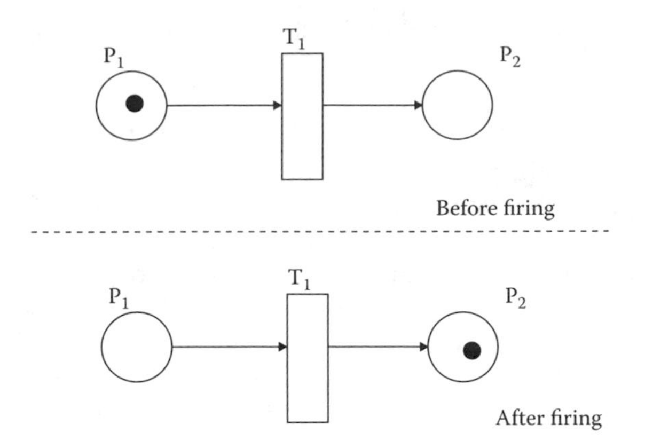

FIGURE 3.9
Petri nets firing rule.

mathematical in nature, techniques for optimization and formal program proving can be employed. But petri nets can be overkill if the system is too simple. Similarly, if the system is highly complex, timing can become obscured.

The petri net is also a powerful tool that can be used for deadlock and race condition identification.

Are there other kinds of petri nets?

The model described herein is just one of a variety of available models. For example, there are timed petri nets, which enable synchronization of firings; colored petri nets, which allow labeled data to propagate through the net; and even timed-colored petri nets, which embody both features.

Are their drawbacks to the use of formal methods?

Formal methods have two limitations that are of special interest to embedded system developers. First, although formalism is often used in pursuit of absolute correctness and safety, it can guarantee neither. Second, formal techniques do not yet offer good ways to reason about alternative designs or architectures.

Correctness and safety are two of the original motivating factors driving adoption of formal methods. Nuclear, defense, and aerospace regulators in

TABLE 3.3
Transition Table for Petri Net Shown in Figure 3.9

	P_1	P_2
Before firing	1	0
After firing	0	1

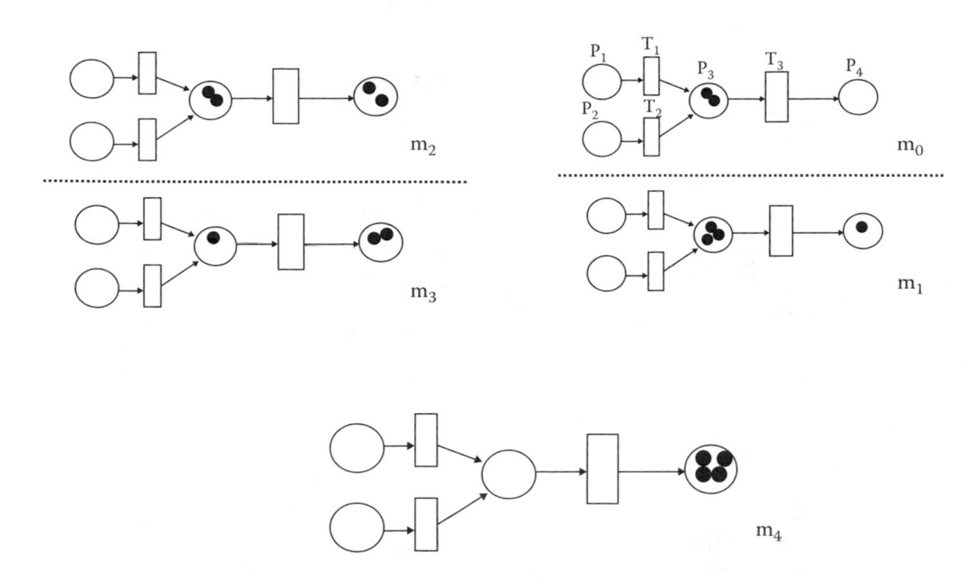

FIGURE 3.10
A slightly more complex petri net.

several countries now mandate or strongly suggest use of formal methods for safety-critical systems. This environment has driven an emphasis on safety-oriented applications of formal methods in the literature. Some researchers emphasize the "correctness" properties of particular mathematical approaches, without clarifying that mathematical correctness in the development process might not translate into real-world correctness in the finished system. After all, it is only the specification that must be produced and proven at this point, not the software product itself.

Formal software specifications must be converted to a design, and later, to a conventional implementation language. This translation process is subject to all of the potential pitfalls of any programming effort. For this reason, testing is just as important when using formal requirement methods as when using traditional ones. Formal verification is also subject to many of the same limitations as traditional testing, namely, testing cannot prove the absence of bugs, only their presence.

TABLE 3.4
Transition Table for Petri Net Shown in Figure 3.9

	P_1	P_2	P_3	P_4
m_0	1	1	2	0
m_1	0	0	3	1
m_2	0	0	2	2
m_3	0	0	1	3
m_4	0	0	0	4

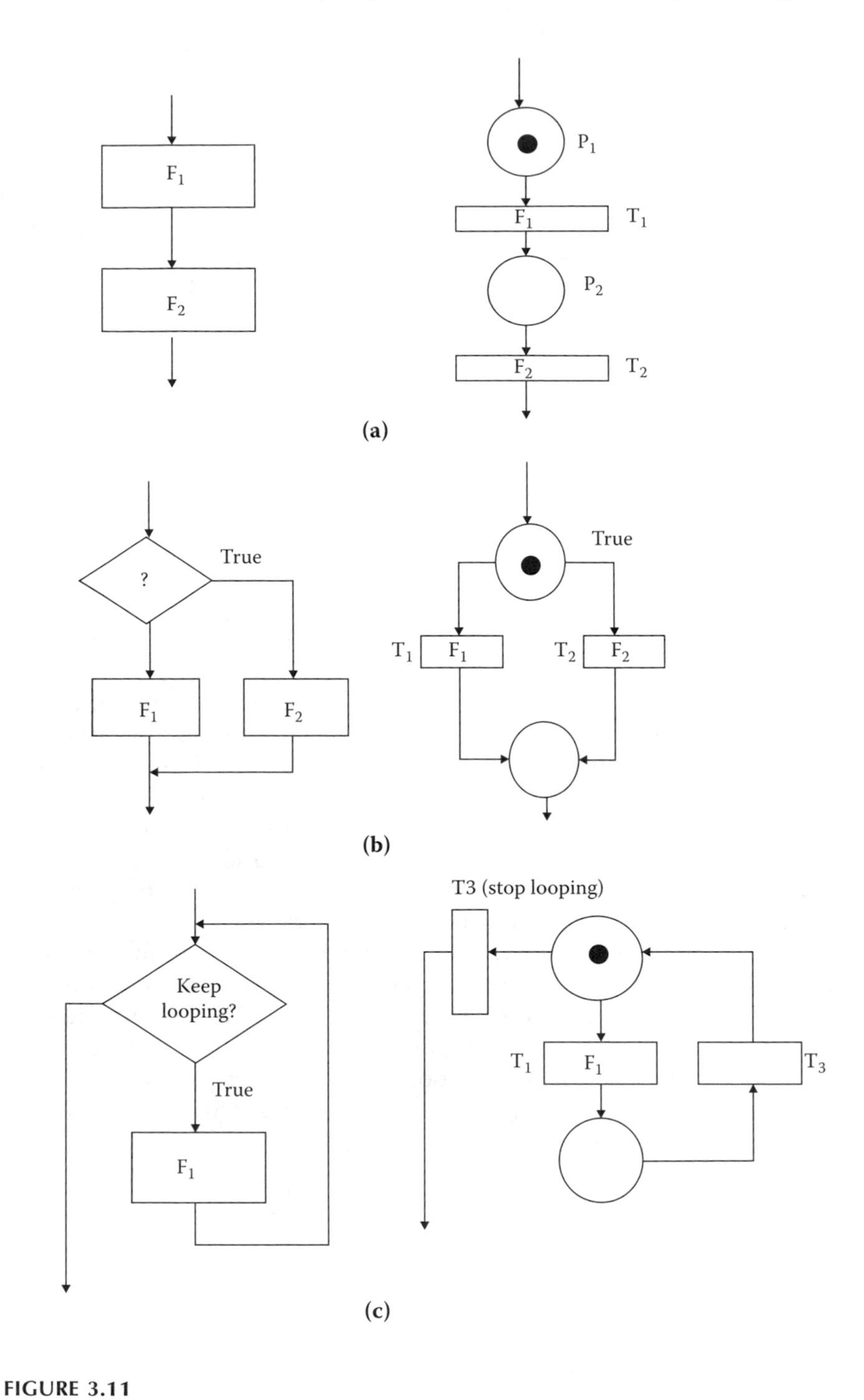

FIGURE 3.11
Flowchart equivalence to petri net configurations: (a) sequence, (b) conditional branch, and (c) `while` loop.

What is structured analysis and structured design?

Structured analysis and structured design (SASD) have evolved over almost 30 years and are widely used in many embedded applications — probably because the techniques are closely associated with the programming languages with which they co-evolved (Fortran and C) and in which many image processing applications are written. Structured methods appear in many forms but the de facto standard is Coad and Yourdon's Modern Structured Analysis [Coad and Yourdon 1991].

Coad and Yourdon's Modern Structured Analysis uses three viewpoints to describe a system: (1) an environmental model, (2) a behavioral model, and (3) an implementation model. The elements of each model are shown in Figure 3.12.

The environmental model embodies the analysis aspect of SASD and consists of a context diagram and an event list. The purpose of the environmental model is to model the system at a high level of abstraction.

The behavioral model embodies the design aspect of SASD as a series of dataflow diagrams (DFDs), entity relationship diagrams (ERDs), process specifications, state transition diagrams, and a data dictionary. Using various combinations of these tools, the designer models the processes, functions, and flows of the system in detail.

Finally, in the implementation model the developer uses a selection of structure charts, natural language, and pseudo-code to describe the system to a level that can be readily translated to code.

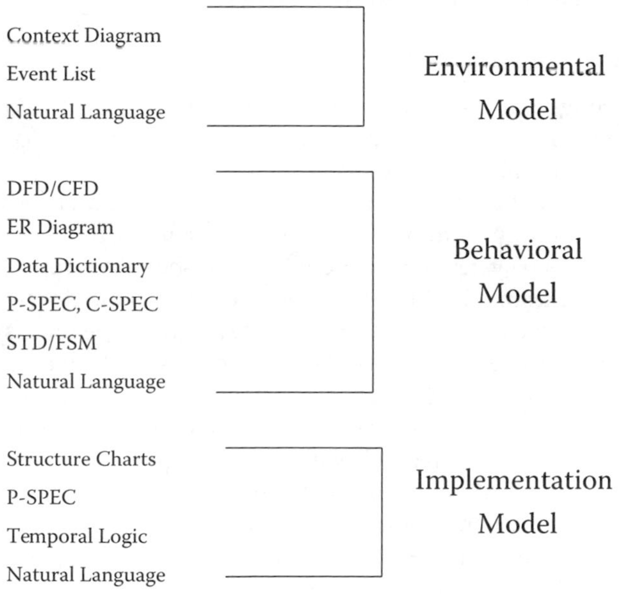

FIGURE 3.12
Elements of structured analysis and design.

What is SA?

SA is a requirements capture technique that tries to overcome the problems of classical analysis using graphical tools and a top-down, functional decomposition method to define system requirements. SA deals only with aspects of analysis that can be structured — the functional specifications and the user interface.

SA is used to model a system's context (where inputs come from and where outputs go), processes (what functions the system performs, how the functions interact, how inputs are transformed to outputs), and content (the data the system needs to perform its functions).

SAs seeks to overcome the problems inherent in analysis through:

- maintainability of the target document
- use of an effective method of partitioning
- use of graphics
- building a logical model of the system for the user before implementation
- reduction of ambiguity and redundancy

What are the main artifacts of SA?

The target document for SA is called the structured specification. It consists of a system context diagram, an integrated set of DFDs showing the decomposition and interconnectivity of components, and an event list to represent the set of events that drive the system.

Can you give an example of SA?

Consider the baggage inspection system previously introduced in its operational mode (calibration and diagnostic modes are ignored for simplicity). Figure 3.13 depicts the context diagram.

The diagram depicts the system's major parts — camera, product detector, conveyor controller system, and reject mechanism. Solid arcs indicate the flow of data between system components. The dashed lines represent the flow of control information. In the example, the only data flow involves the transmission of the captured image to the baggage inspection system.

In the example, the event list consists of the **new_bag_event**, which indicates the detection of the next suitcase on the line; **accept**, which indicates that the suitcase has passed inspection and causes a signal to be sent to the conveyor controller; and **reject**, which causes a signal to be sent that directs the conveyor to move the bag into a separate bin. A piece of luggage may be rejected for one reason or another such as a security threat or if the bag is too large. The rejection mechanism automatically causes the next bag to be moved along by the conveyor system.

It should be reiterated that this context diagram is not complete owing to the omission of the calibration and diagnostic modes. While the intent here is not to provide a complete system design, missing functionality is more

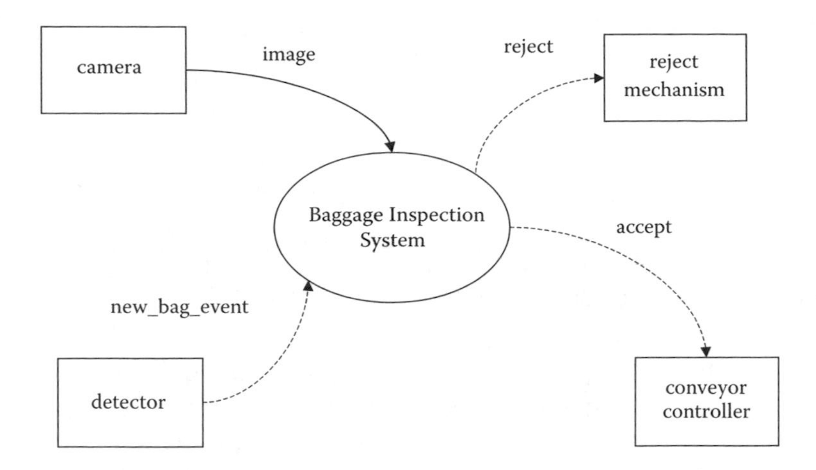

FIGURE 3.13
Context diagram for the baggage inspection system.

easily identified during the requirements elicitation process if some form of graphical aid, such as the context diagram, is available. In the case of OOA, a use case diagram will be helpful.

What does "object-oriented" mean?

"Object-oriented" defines a paradigm for describing system behavior in which entities are attributed characteristics (called attributes), have operations that can be associated with them (called methods), and can interact with each other through messaging or shared data structures. Object-oriented programming is believed by many to lead to more maintainable, understandable, and extendable systems. Some believe that object-oriented software engineering is easier to learn and master than its counterpart, procedural-oriented (or imperative) engineering. Languages written to support the object-oriented paradigm are called object-oriented languages.

What is object-oriented requirements analysis?

As an alternative to the SA approach to developing software requirements for the baggage inspection system, consider using an object-oriented approach. There are various "flavors" of object-oriented approaches, each using their own toolsets. In the approach developed here, the system specification begins with the representation of externally accessible functionality as use cases.

When is it appropriate to use OOA vs. SA?

Both SA and OOA can use the tools and techniques for modeling as previously described. However, there are major differences between the two techniques. SA describes the system from a functional perspective and separates dataflow from the functions that transform them, while OOA describes the system from the perspective of encapsulated entities that possess both function and form.

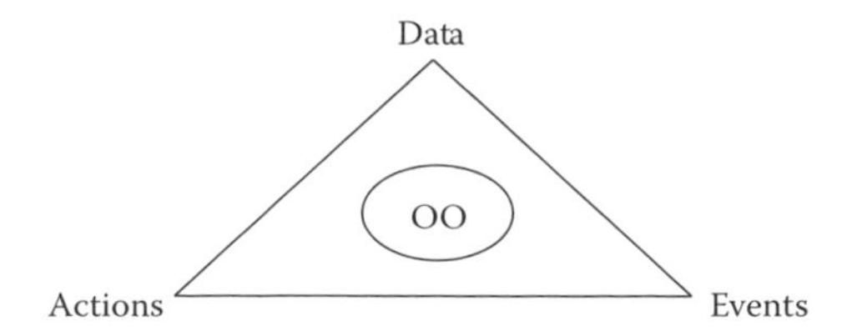

FIGURE 3.14
A project's applicability to either OOA or SA according to system focus.

Additionally, object-oriented models include inheritance while SA does not. While SA has a definite hierarchical structure, this is a hierarchy of composition rather than heredity. This shortcoming leads to difficulties in maintaining and extending both the specification and design such as in the case of changes in the baggage inspection system example.

The purpose of this discussion is not to dismiss SA, or even to conclude that it is better than OOA in all cases. An overriding indicator of suitability of OOA vs. SA is the nature of the application. To understand this, consider the vertices of the triangle in Figure 3.14 as representing three distinct viewpoints of a system: data, actions, and events.

Events represent stimuli and responses such as measurements in process control systems, as in the case study. Actions are rules that are followed in complex algorithms, such as "binarize," "threshold," and "classify." The majority of early computer systems were focused on one, or at most two, of these vertices. For example, early, non-real-time systems were data and action intensive but did not encounter much in the way of stimuli and response.

3.6 Requirements Documentation

What is the role of the SRS?

The SRS document is the official statement of what is required of the system developers. The SRS should include both a definition and a specification of requirements. However, the SRS is *not* a design document. As much as possible, it should be a set of *what* the system should do rather than *how* it should do it. Unfortunately, the SRS usually contains some design specifications, which has the tendency to hamstring the designers.

Who uses the requirements documents?

A variety of stakeholders uses the software requirements throughout the software life cycle. Stakeholders include customers (these might be external customers or internal customers such as the marketing department), managers, developers, testers, and those who maintain the system. Each stakeholder

TABLE 3.5

Software Stakeholders and Their Uses of the SRS

Stakeholder	Use
Customers	Express how their needs can be met. They continue to do this throughout the process as their perceptions of their own needs change.
Managers	Bid on the system and then control the software production process.
Developers	Create a software design that will meet the requirements.
Test engineers	A basis for verifying that the system performs as required.
Maintenance engineers	Understand what the system was intended to do as it evolves over time.

has a different perspective on and use for the SRS. Various stakeholders and their uses for the SRS are summarized in Table 3.5.

How do I organize the requirements document?

There are many ways to organize the SRS; however, IEEE Standard 830-1998 Recommended Practice for Software Requirements Specifications (SRS) provides a template of what an SRS should look like.

The SRS starts with "boilerplate" frontmatter, which usually acts as a preamble to the product. For example, in Appendix A, most of Sections 1 and 2 introduce the wet well control system terminology and operating environment.

Following the boilerplate material, the functionality of the system is described in one or more styles (or "views"). IEEE 830 recommends that the views include some combination of decomposition, dependency, interface, and detail descriptions. Together with the boilerplate frontmatter, these form a standard template for SRSs, which is depicted in Figure 3.15.

Sections 1 and 2 of Appendix A are self-evident; they provide frontmatter and introductory material for the SRS. The remainder of the SRS is devoted to the four description sections.

An outline of some specific requirements, or work breakdown structure (WBS), for the baggage inspection system, in IEEE 830 format, is given in Figure 3.16.

The section headings can be decomposed further using a technique such as SA or OOA. A more complete example using the wet well control system can be found in Appendix A.

How do you represent specific requirements in the SRS?

The IEEE 830 standard provides for several alternative means to represent the requirements specifications, aside from a function perspective. In particular, the software requirements can be organized by

- functional mode ("operational," "diagnostic," "calibration")
- user class ("operator," "diagnostic")
- object
- feature (what the system provides to the user)

1. Introduction
    1.1 Purpose
    1.2 Scope
    1.3 Definitions and Acronyms
    1.4 References
    1.5 Overview
2. Overall description
    2.1 Product perspective
    2.2 Product functions
    2.3 User characteristics
    2.4 Constraints
    2.5 Assumptions and dependencies
3. Specific Requirements
Appendices
Index

FIGURE 3.15
Recommended Table of Contents for an SRS from IEEE Standard 830–1998.

3. Functional Requirements
3.1 Calibration mode
3.2 Operational mode
3.2.1 Initialization
3.2.2 Normal operation
3.2.2.1 Image capture
3.2.2.2 Image error correction
3.2.2.2.1 Position error reduction
3.2.2.2.2 Noise error reduction
3.2.2.3 Captured image analysis
3.2.2.4 Conveyor system control
3.2.2.5 Reject mechanism control
3.2.2.6 Error handling
3.3 Diagnostic Mode
4. Nonfunctional Requirements

FIGURE 3.16
Some specific requirements for the baggage inspection system.

- stimulus (sensor 1, 2, etc.)
- functional hierarchy
- mixed (a combination of two or more of the above; e.g., in Appendix A, the requirements include feature- and stimulus-based descriptions)

What is requirements traceability?

Requirements traceability is concerned with the relationships between requirements, their sources, and the system design. Requirements can be linked to the source, to other requirements, and to design elements.

Source traceability links requirements to the stakeholders who proposed these requirements. Requirements traceability links between dependent requirements. Design traceability links from the requirements to the design.

What role does traceability help in requirements documentation?

During the requirements engineering process, the requirements must be identified to assist in planning for the many changes that will occur throughout the software life cycle. Traceability is also an important factor in conducting an impact analysis to determine the effort needed to make such changes.

What are traceability policies?

Traceability policies determine the amount of information about requirements relationships that need to be maintained. There are a number of open source and proprietary CASE tools that can help improve requirements traceability.

What does a traceability matrix look like?

One type of traceability matrix is shown in Table 3.6. Requirements identification numbers label both the rows and columns. An "R" is placed in a corresponding cell if the requirement in that row references the requirement in that column. A "U" corresponds to an actual use dependency between the two requirements.

TABLE 3.6

A Sample Traceability Matrix

Requirement Identification	1.1	1.2	1.3	2.1	2.2	2.3	3.1	3.2
1.1		U	R					
1.2			U			R		U
1.3	R			R				
2.1			R		U			U
2.2								U
2.3		R		U				
3.1								R
3.2							R	

3.7 Recommendations on Requirements

Is there a preferred modeling technique for an SRS?

It is risky to prescribe a preferred technique because it is well known that there is no "silver bullet" when it comes to software specification and design and each system should be considered on its own merits.

Nevertheless, regardless of approach, any SRS should incorporate the following best practices:

- Use consistent modeling approaches and techniques throughout the specification; for example, top-down decomposition, SA or OOA.
- Separate operational specification from descriptive behavior.
- Use consistent levels of abstraction within models and conformance between levels of refinement across models.
- Model nonfunctional requirements as a part of the specification models, in particular timing properties.
- Omit hardware and software assignment in the specification (another aspect of design rather than specification).

Are there special challenges when engineers specify software systems?

The preceding discussions illustrate some of the challenges (in fact, one might consider them "habits") encountered by engineers specifying software systems. The challenges include:

- Mixing of operational and descriptive specifications.
- Combining low-level hardware functionality and high-level systems and software functionality in the same functional level.
- Omission of timing information.

Whatever approach is used in organizing the SRS, the IEEE 830 standard describes the characteristics of good requirements. They are as follows:

Correct — The SRS should correctly describe the system behavior. Obviously, it is unhelpful if the SRS incorrectly defines the system or somehow involves unreasonable expectations such as defying the laws of physics.

Unambiguous — An unambiguous SRS is one that is clear and not subject to different interpretations. Using appropriate language can help avoid ambiguity.

Complete — An SRS is complete if it completely describes the desired behavior. Ordinarily, the note "TBD," that is "to be defined (later)" is unacceptable in a requirements document. IEEE 830 sets out some exceptions to this rule.

Consistent — One requirement must not contradict another. Consistency checking is discussed in Chapter 6.

Ranked — An SRS must be ranked for importance and/or stability. Not every requirement is as critical as another. By ranking the requirements, designers will find guidance in making tradeoff decisions.

Verifiable — Any requirement that cannot be verified is a requirement that cannot be shown to have been met.

Modifiable — The requirements need to be written in such a way so as to be easy to change. In many ways, this approach is similar to the information hiding principle.

Traceable — The SRS must be traceable because the requirements provide the starting point for the traceability chain. Approaches to traceability and its benefits have been mentioned at length.

How can I rank requirements?

For example, NASA uses four levels of requirements. Level 1 requirements are mission level requirements. These requirements are very high level, and rarely change. Level 2 requirements are high level, and will change minimally. Level 3 requirements are mid-level requirements derived from Level 2. Each Level 2 requirement traces to one or more Level 3 requirement. Most contracts bid at this level of detail. Finally, Level 4 requirements are very detailed and are used to design and code the system.

I sometimes advocate a three-level taxonomy of requirements — mandatory, desirable, and optional. Mandatory requirements cannot be sacrificed. Desirable requirements are important but could be sacrificed if necessary to meet schedule or budget. Optional requirements would be nice to have, but are readily sacrificed.

Why does ranking requirements do for me?

Ranking requirements is quite helpful when tradeoffs need to be made. For example, if time or work force is limited, then place the focus on the higher ranked requirements. The same principle holds for testing; if testing time is limited, then it can be focused on the requirements pertaining to the higher-level requirements.

What wording is appropriate in requirements specifications?

To meet these criteria and to write clear requirements documentation, there are several best practices that the requirements engineer can follow. They are as follows:

- Invent and use a standard format for all requirements.
- Use language in a consistent way.
- Use "shall" for mandatory requirements.

- Use "should" for desirable requirements.
- Use text highlighting to identify key parts of the requirement.
- Avoid the use of technical language unless it is warranted.

How do I recognize bad requirements?

To illustrate, consider the following bad requirements.

"The systems shall be completely reliable."

"The system shall be modular."

"The system shall be maintainable."

"The system will be fast."

"Errors shall be less than 99%."

These requirements are bad for a number of reasons. None is verifiable; for example, how is "reliability" supposed to be measured? Even the last requirement is vague. What does less than 99% mean?

What do some good requirements look like?

Consider the following requirements:

"Response times for all Level 1 actions will be less than 100 ms."

"The cyclomatic complexity of each module shall be in the range of 10 to 40."

"95% of the transactions shall be processed in less than 1 s."

"An operator shall not have to wait for the transaction to complete."

"MTBF shall be 100 hours of continuous operation."

These requirements are "better" versions of the preceding ones. Each is measurable because each makes some attempt to quantify the qualities that are desired. For example, cyclomatic complexity is a measure of modularity, MTBF is a measure of failures, and processing time is a measure of speed. Nevertheless, these improved requirements could stand some refinement based on the context of requirements specification as a whole.

Also notice that most of the requirements in Appendix A read as "the system shall…". This requirements specification closely follows the 830 guidelines. The requirements are not ranked in this case, however, because it is a very small system with virtually all mandatory functionality. We could always add various desirable and optional features and rank the SRS accordingly.

What is requirements triage?

The following is good summary advice from requirements engineering specialist Al Davis. When dealing with the swirl of issues involved with requirements engineering, he suggests that you:

Maintain a list of requirements.
Record necessary interdependencies.
Annotate requirements by effort.
Annotate requirements by relative importance.
Perform triage overtly (involve stakeholders)

- Customers
- Developers
- Financial representatives.

Base decisions on more than mechanics.
Establish a teamwork mentality.
Manage by probabilities of completion not absolutes.
Understand the optimistic, pessimistic, and realistic approaches.
Plan more than one release at a time.
Replan before every new release.
Don't be intimidated into a solution.
Find a solution before you proceed.
Remember that prediction is impossible [Davis 2003].

This is good advice to keep in mind.

What is requirements validation?

Requirements validation is tantamount to asking the question "Am I building the right software?" Too often, software engineers deliver a system that conforms to the SRS only to discover that it is not what the customer really wanted.

Requirements error costs are high, and therefore validation is very important. Fixing a requirements error after delivery may cost up to 100 times the cost of fixing an implementation error.

How are requirements validated?

There are number of ways of checking SRSs for conformance to the IEEE 830 best practices and for validity. These approaches include:

requirements reviews
systematic manual analysis of the requirements
prototyping
using an executable model of the system to check requirements

test-case generation
developing tests for requirements to check testability
automated consistency analysis
checking the consistency of a structured requirements description

Can requirements specifications be automatically checked for quality?

Automated checking is the most desirable and the least likely because of the context sensitivity of natural languages, and the impossibility of verifying such things as requirements completeness. However, simple tools can been developed to perform spelling and grammar checking, flagging of key words that may be ambiguous (e.g., "fast," "reliable"), identification of missing requirements (e.g., search for the phrase "to be determined"), and overly complex sentences (like this one, which can indicate unclear requirements).

Are there tools out there already?

There is a free downloadable tool from NASA, the Automated Requirement Measurement (ARM) tool [NASA 2006]. This tool measures the "goodness" of a requirements specification on two levels, macro and micro. The macro level indicators include:

- Size of requirements
- Text structure
- Specification depth
- Readability

Micro level indicators deal with specific language utilization, such as the number of instances of "shall" vs. "should."

How does the tool measure the size of requirements?

Size of requirements can be measured in terms of lines of text, number of paragraphs, and certain ratios of key words, which can indicate level of detail or conciseness.

How does the tool measure text structure?

Text structure is a kind of topological feature of the document that indicates the number of statement identifiers at each hierarchical level. High-level requirements rarely have numbered statements below a depth of 4 (e.g., 3.2.1.5). Well-organized documents have a pyramidal structure to the requirements. An hourglass structure means too many administrative details. A diamond structure indicate subjects introduced at higher levels were addressed at different levels of detail (see Figure 3.17).

How does the tool measure specification depth?

Specification depth measures the number of imperatives found at each document level.

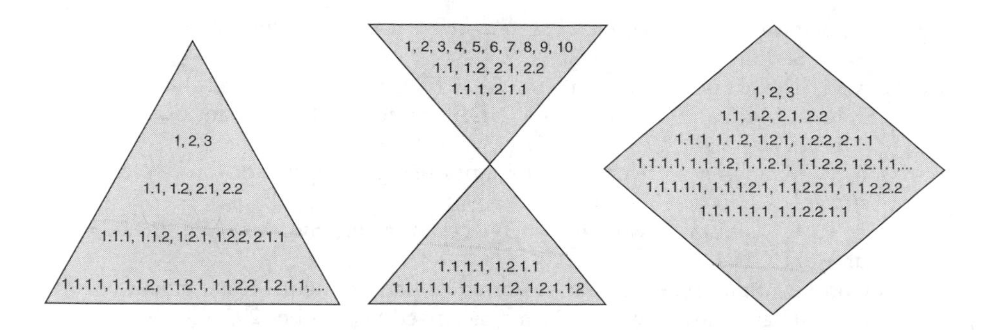

FIGURE 3.17
Text structure: (a) desirable, (b) missing some mid-level details, (c) not enough low-level detail.

The number of lower level items that are introduced at a higher level by an imperative followed by a continuance are also counted. The ratio of specification depth/total lines of text indicates conciseness of the SRS.

How are reading statistics useful in assessing an SRS?

The readability statistics rely on traditional measures of human writing level such as:

- Flesch Reading Ease Index — the number of syllables per word and words per sentence.
- Flesch-Kincaid Grade Level Index — Flesch score converted to a grade level (standard writing is about 7th or 8th grade).
- Coleman-Liau Grade Level Index — uses word length in characters and sentence length in words to determine grade level.
- Bormuth Grade Level Index — also uses word length in characters and sentence length in words to determine grade level.

Many of these indicators are calculated by conventional word processors. Organizations can choose an appropriate level for any of these metrics as a standard for software documentation.

3.8 Further Reading

Akao, Y. (Ed.). *Quality Function Deployment*, Productivity Press, Cambridge, MA, 1990.

Beyer, H. and Holtzblatt, K., Apprenticing with the customer: a collaborative approach to requirements definition, *Commun. ACM*, 38(5), 45–52, 1995.

Coad, P. and E. Yourdon, *Object-Oriented Analysis,* 2nd ed., Yourdon Press, Upper Saddle River, NJ, 1990.

Davis, A., *Software Requirements: Objects, Functions and States,* 2nd ed., Prentice-Hall, Upper Saddle River, NJ, 1993.

Davis, A.M., The art of requirements triage, *IEEE Computer,* 36(3), 42–49, 2003.

Haag, S., Raja, M.K., and Schkade, L.L., Quality function deployment usage in software development, *Commun. ACM,* 39(1), 41–49, 1996.

Karlsson, J., Managing software requirements using quality function deployment, *Software Q. J.,* 6(4), 311–326, 1997.

Laplante, P. A., *Software Engineering for Image Processing Systems,* CRC Press, Boca Raton, FL, 2004.

NASA Goddard Spaceflight Center, Automated Requirement Measurement (ARM) Tool, satc.gsfc.nasa.gov/tools/arm/, accessed September 29, 2006.

Nuseibeh, B. and Easterbrook, S., Requirements engineering: a roadmap, *Proc. Intl. Conf. Software Eng.,* Limerick, Ireland, June 2000, 35–46.

Sommerville, I., *Software Engineering,* 7th ed., Addison-Wesley, Boston, MA, 2005.

Wilson, W.M., "Writing Effective Requirements Specifications," presented at the Software Technology Conference, Salt Lake City, Utah, April 1997.

Zave, P., Classification of research efforts in requirements engineering, *ACM Comp. Surveys,* 29(4), 315–321, 1997.

4

Designing Software

Outline

- Software design concepts
- Software design modeling
- Pattern-based design
- Design documentation

4.1 Introduction

Mature engineering disciplines have handbooks that describe successful solutions to known problems. For example, automobile designers don't design cars using the laws of physics; they adapt adequate solutions from the handbook known to work well enough.

The extra few percent of performance available by starting from scratch typically is not worth the cost.

Software engineering has been criticized for not having the same kind of underlying rigor as other engineering disciplines. But while it may be true that there are few formulaic principles, there are fundamental rules that form the basis of sound software engineering practice. These rules can form the basis of the handbook for software engineering.

If software is to become an engineering discipline, successful practices must be systematically documented and widely disseminated. The following sections describe the most general and prevalent of these concepts and practices for software design.

4.2 Software Design Concepts

What is software design?

Software design involves identifying the components of the software design, their inner workings, and their interfaces from the SRS. The principle artifact of this activity is the software design specification (SDS), which is also referred to as a software design description (SDD).

What are the principal activities of software design?

During the design activity, the engineer must design the software architecture, which involves any set of the following tasks.

- Performing hardware/software tradeoff analysis.
- Designing interfaces to external components (hardware, software, and user interfaces).
- Designing interfaces between components.
- Making the determination between centralized or distributed processing schemes.
- Determining concurrency of execution.
- Designing control strategies.
- Determining data storage, maintenance, and allocation strategy.
- Designing databases, structures, and handling routines.
- Designing the startup and shutdown processing.
- Designing algorithms and functional processing.
- Designing error processing and error message handling.
- Conducting performance analyses.
- Specifying physical location of components and data.
- Creating documentation for the system
- Conducting internal reviews.
- Developing the detailed design for the components identified in the software architecture.
- Documenting the software architecture in the form of the SDS.
- Presenting the detail design information at a formal design review.

This is an intimidating set of tasks that is further complicated as many of these must occur in parallel or be iterated several times. Moreover, because clearly more than one individual must be involved in these activities, problems of working on a project team come into play.

There are no rules of thumb per se for conducting these tasks. Instead, it takes many years of practice, experience, learning from the experience of others, and good judgment to guide the software engineer through this maze of tasks.

4.2.1 Basic Software Engineering Principles

How do rigor and formality enter into software engineering?

Because software development is a creative activity, there is an inherent tendency toward informal *ad hoc* techniques in software specification, design, and coding. But the informal approach is contrary to good software engineering practices.

Rigor in software engineering requires the use of mathematical techniques. Formality is a higher form of rigor in which precise engineering approaches are used. For example, imaging systems require the use of rigorous mathematical specification in the description of image acquisition, filtering, enhancement, etc. But the existence of mathematical equations in the requirements or design does not imply an overall formal software engineering approach. In the case of the baggage inspection system, formality further requires that there be an underlying algorithmic approach to the specification, design, coding, and documentation of the software.

What is separation of concerns?

Separation of concerns is a kind of divide and conquer strategy that software engineers use. There are various ways in which separation of concerns can be achieved. In terms of software design and coding, it is found in modularization of code and in object-oriented design. There may be separation in time; for example, developing a schedule for a collection of periodic computing tasks with different periods.

Yet another way of separating concerns is in dealing with qualities. For example, it may be helpful to address the fault-tolerance of a system while ignoring other qualities. However, it must be remembered that many of the qualities of software are interrelated, and it is generally impossible to affect one without affecting the other, possible adversely.

Can modular design lead to separation of concerns?

Some separation of concerns can be achieved in software through modular design. Modular design, first proposed by Parnas [1972], involves the decomposition of software behavior in encapsulated software units. Separation of concerns can be achieved in either object-oriented or procedurally oriented programming languages.

Modularity is achieved by grouping together logically related elements, such as statements, procedures, variable declarations, object attributes, and so on in increasingly greater levels of detail (Figure 4.1).

The main objectives in seeking modularity are to foster high cohesion and low coupling. With respect to the code units, cohesion represents intra-module

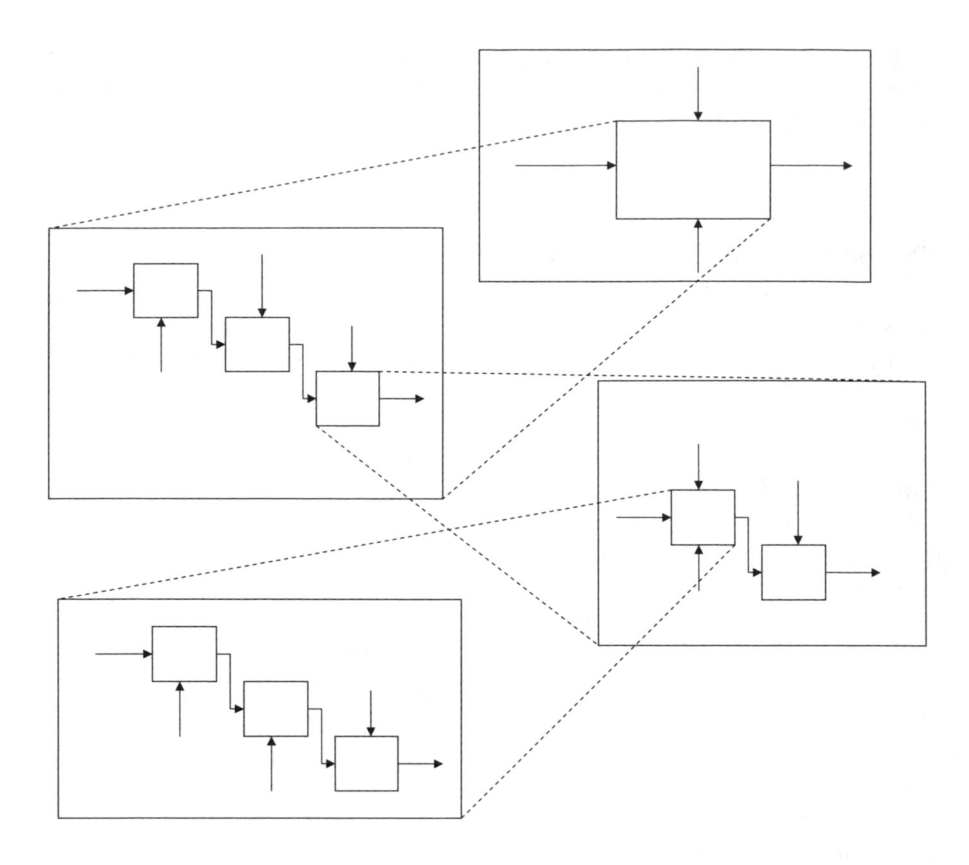

FIGURE 4.1
Modular decomposition of code units. The arrows represent inputs and outputs in the procedural paradigm. In the object-oriented paradigm, they represent method invocations or messages. The boxes represent encapsulated data and procedures in the procedural paradigm. In the object-oriented paradigm they represent classes.

connectivity and coupling represents inter-module connectivity. Coupling and cohesion can be illustrated informally as in Figure 4.2, which shows software structures with high cohesion and low coupling (Figure 4.2a) and low cohesion and high coupling (Figure 4.2b). The inside squares represent statements or data, and the arcs indicate functional dependency.

What is cohesion?

Cohesion relates to the relationship of the elements of a module. Constantine and Yourdon identified seven levels of cohesion in order of strength [Pressman 2005]:

Coincidental — parts of the module are not related but are simply bundled into a single module.

Logical — parts that perform similar tasks are put together in a module.

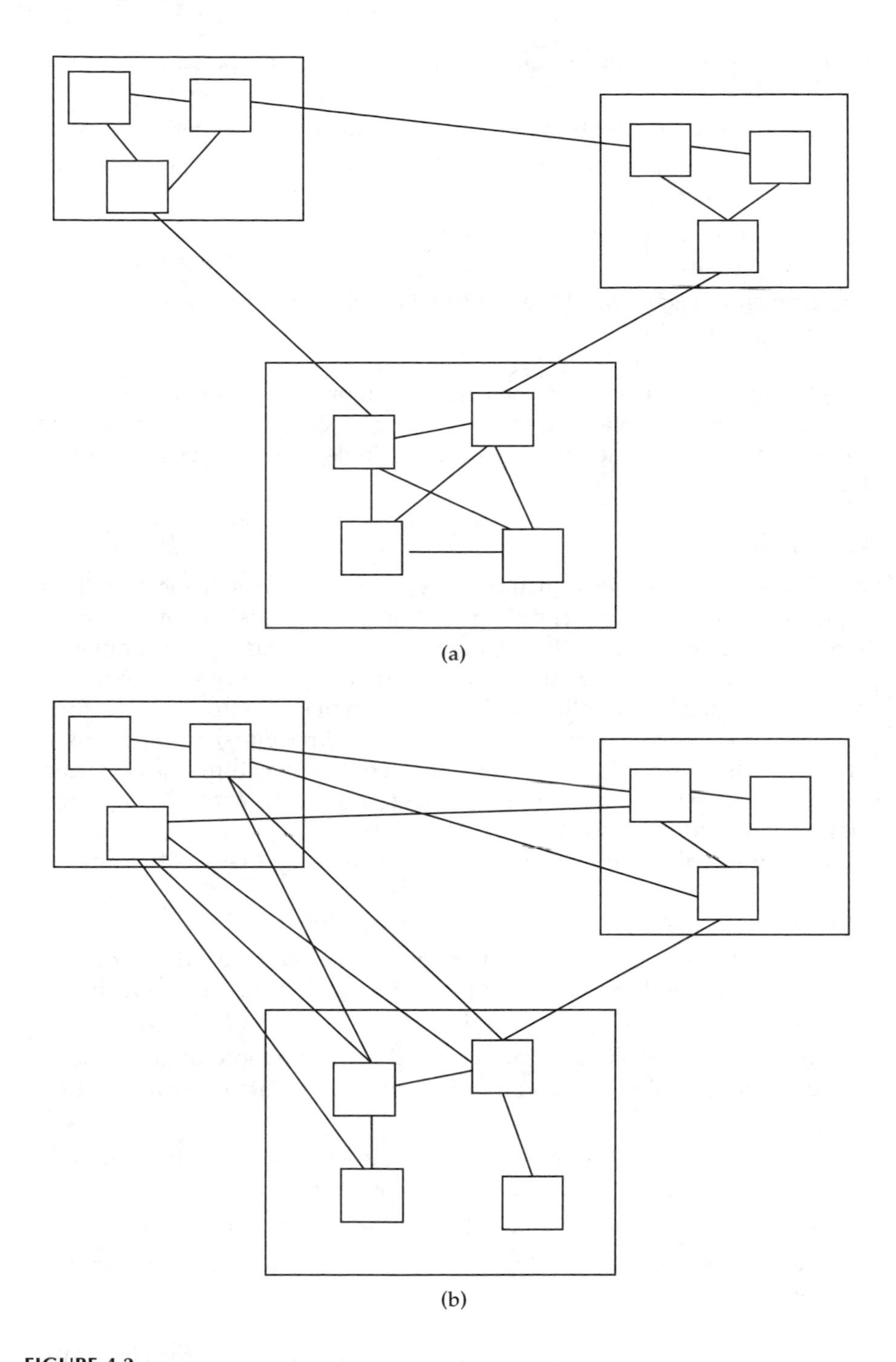

FIGURE 4.2
Software structures with high cohesion and low coupling (a) and low cohesion and high coupling (b). The inside squares represent statements or data, and the arcs indicate functional dependency.

Temporal — tasks that execute within the same time span are brought together.

Procedural — the elements of a module make up a single control sequence.

Communicational — all elements of a module act on the same area of a data structure.

Sequential — the output of one part of a module serves as input for another part.

Functional — each part of the module is necessary for the execution of a single function.

High cohesion implies that each module represents a single part of the problem solution. Therefore, if the system ever needs modification, then the part that needs to be modified exists in a single place, making it easier to change.

What is coupling?

Coupling relates to the relationships between the modules themselves. There is great benefit in reducing coupling so that changes made to one code unit do not propagate to others; that is, they are hidden. This principle of "information hiding," also known as Parnas partitioning, is the cornerstone of all software design. Low coupling limits the effects of errors in a module (lower "ripple effect") and reduces the likelihood of data integrity problems. In some cases, however, high coupling due to control structures may be necessary. For example, in most graphical user interfaces control coupling is unavoidable, and indeed desirable.

Coupling has also been characterized in increasing levels as follows:

No direct coupling — all modules are completely unrelated.

Data — when all arguments are homogeneous data items; that is, every argument is either a simple argument or data structure in which all elements are used by the called module.

Stamp — when a data structure is passed from one module to another, but that module operates on only some of the data elements of the structure.

Control — one module passes an element of control to another; that is, one module explicitly controls the logic of the other.

Common — if two modules both have access to the same global data.

Content — one module directly references the contents of another [Parnas 1972].

To further illustrate both coupling and cohesion, consider the class structure for a widely used commercial imaging application program interface (API) package, depicted in Figure 4.3.

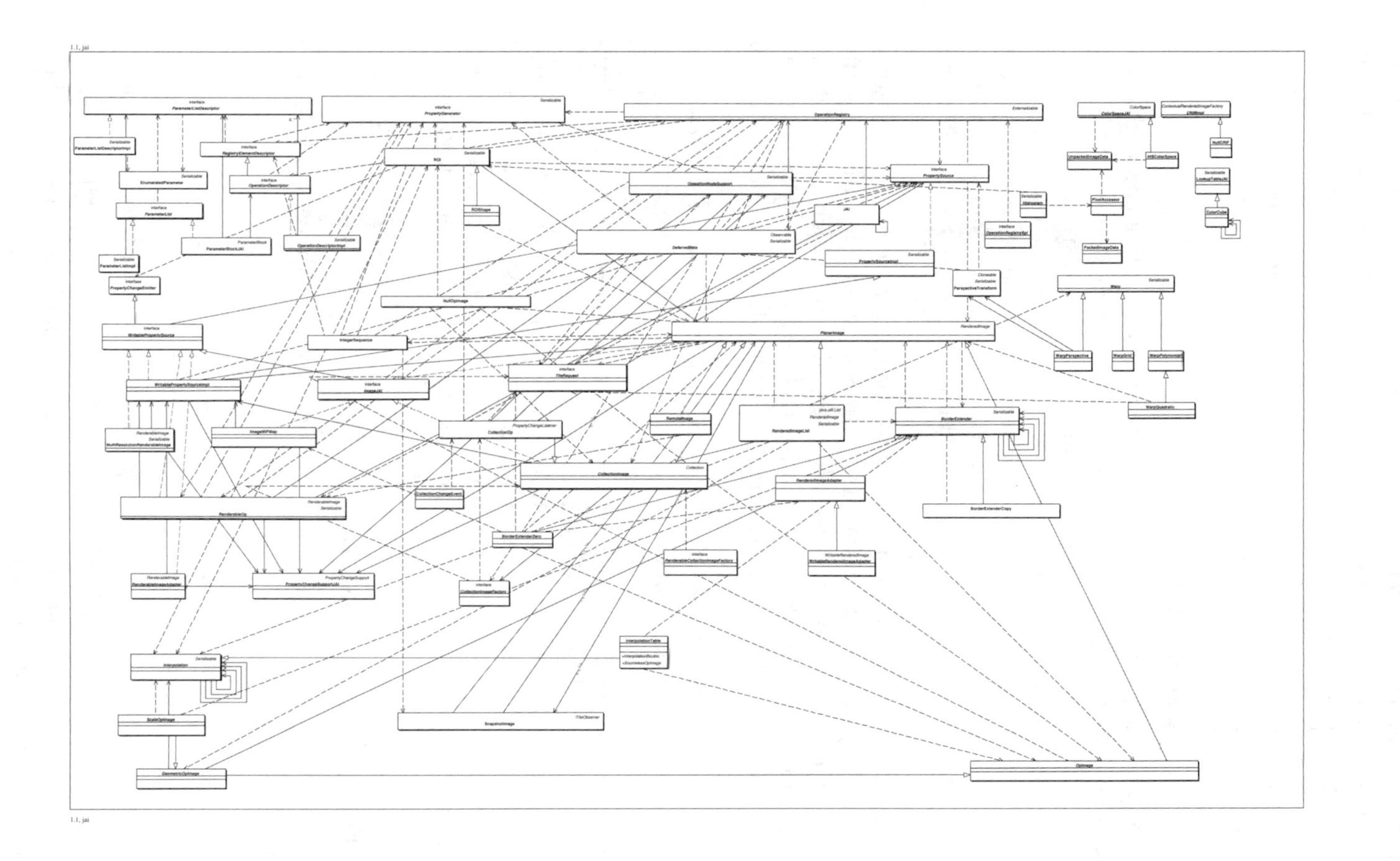

FIGURE 4.3
The class structure for the API of a widely deployed imaging package.

The class diagram was obtained through design recovery. The class names are not readable in the figure, but it is not the intention to identify the software. Rather, the point is to illustrate the fact that there is a high level of coupling and low cohesion in the structure. This design would benefit from refactoring; that is, performing a behavior preserving code transformation, which would achieve higher cohesion and lower coupling.

What is Parnas partitioning?

Software partitioning into software units with low coupling and high cohesion can be achieved through the principle of information hiding. In this technique, a list of difficult design decisions or things that are likely to change is prepared. Code units are then designated to "hide" the eventual implementation of each design decision or feature from the rest of the system. Thus, only the function of the code units is visible to other modules, not the method of implementation. Changes in these code units are therefore not likely to affect the rest of the system.

This form of functional decomposition is based on the notion that some aspects of a system are fundamental, whereas others are arbitrary and likely to change. Moreover, those arbitrary things, which are likely to change, contain "information." Arbitrary facts are hard to remember and usually require lengthier descriptions; therefore, they are the sources of complexity.

Parnas partitioning "hides" the implementation details of software features, design decisions, low-level drivers, etc., in order to limit the scope of impact of future changes or corrections. By partitioning things likely to change, only that module need be touched when a change is required without the need to modify unaffected code.

This technique is particularly applicable and useful in embedded systems. Because they are so directly tied to hardware, it is important to partition and localize each implementation detail with a particular hardware interface. This allows easier future modification due to hardware interface changes and reduces the amount of code affected.

How do I do Parnas partitioning?

The following steps can be used to implement a design that embodies information hiding.

Begin by characterizing the likely changes.

Estimate the probabilities of each type of change.

Organize the software to confine likely changes to a small amount of code.

Provide an "abstract interface" that abstracts from the differences.

Implement "objects;" that is, abstract data types and modules that hide changeable data structures.

These steps reduce coupling and increase module cohesion. Parnas also indicated that although module design is easy to describe in textbooks, it is difficult to achieve. He suggested that extensive practice and examples are needed to illustrate the point correctly [Parnas 1972].

Can you give an example of Parnas partitioning?

Consider a portion of the display function of the baggage inspection system shown in hierarchical form in Figure 4.4. It consists of graphics that must be displayed (for example, a representation of the conveyor system, units moving along it, sensor data, etc.), which are essentially composed of circles and boxes. Different objects can also reside in different display windows. The implementation of circles and boxes is based on the composition of line drawing calls. Thus, line drawing is the most basic hardware-dependent function. Whether the hardware is based on pixel, vector, turtle, or other type of graphics does not matter; only the line drawing routine needs to be changed. Hence, the hardware dependencies have been isolated to a single code unit.

If when designing the software modules, increasing levels of detail are deferred until later, then the software approach is top-down. If, instead, the design detail is dealt with first and then increasing levels of abstraction are used to encapsulate those details, then the approach is bottom-up.

For example, in Figure 4.4, it would be possible to design the software by first describing the characteristics of the various components of the system and the functions that are to be performed on them, such as opening, closing, and sizing windows. Then the window functionality could be decomposed into its constituent parts such as boxes and text. These could be decomposed still further; for example, all boxes consist of lines and so on. The top-down

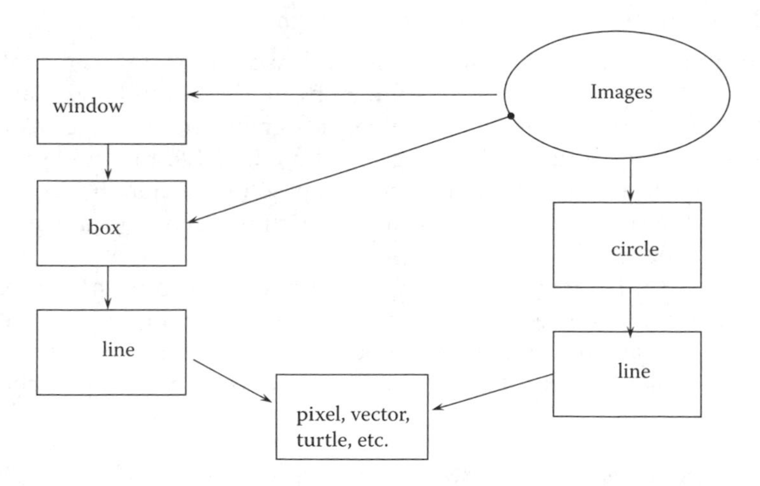

FIGURE 4.4
Parnas partitioning of graphics rendering software.

refinement continues until the lowest level of detail needed for code development has been reached.

Alternatively, it is possible to begin by encapsulating the details of the most volatile part of the system, the hardware implementation of a single line or pixel, into a single code unit. Then working upward, increasing levels of abstraction are created until the system requirements are satisfied. This is a bottom-up approach to design.

Can you do Parnas partitioning in object-oriented design?

Yes. In the object-oriented paradigm, Parnas partitioning is a form of a design technique called protected variation, which is one of the "GRASP" principles (to be discussed in Section 4.4).

How can changes be anticipated in software designs?

It has been mentioned that software products are subject to frequent change either to support new hardware or software requirements or to repair defects. A high maintainability level of the software product is one of the hallmarks of outstanding commercial software.

Engineers know that their systems are frequently subject to changes in hardware, algorithms, and even application. Therefore, these systems must be designed in such a way to facilitate changes without degrading the other desirable properties of the software.

Anticipation of change can be achieved in the software design through appropriate techniques, through the adoption of an appropriate software life-cycle model and associated methodologies, and through appropriate management practices.

How does generality apply to software design?

In solving a problem, the Principle of Generality can be stated as the intent to look for the more general problem that may be hidden behind a specialization of that problem. In an obvious example, designing the baggage inspection system is less general than designing it to be adaptable to a wide range of inspection applications. Although generalized solutions may be more costly in terms of the problem at hand, in the long run the costs of a generalized solution may be worthwhile.

Generality can be achieved through a number of approaches associated with procedural and object-oriented paradigms. For example, in procedural languages, Parnas' information hiding can be used. In object-oriented languages, the Liskov Substitution Principle can be used. Both approaches will be discussed shortly.

How does incrementality manifest in software design?

Incrementality involves a software approach in which progressively larger increments of the desired product are developed. Each increment provides additional functionality, which brings the product closer to the final one.

Each increment also offers an opportunity for demonstration of the product to the customer for the purposes of gathering requirements and refining the look and feel of the product.

4.2.2 Software Architectures

What is a software architecture?

This is a relatively new concept in software engineering. A software architecture is the structure and organization by which modern system components and subsystems interact to form systems. The software architecture is the embodiment of the properties of systems that can best be designed and analyzed at the system level [Kruchten et al. 2006].

The work of Parnas [1972], particularly his notions of information hiding and the object-oriented equivalent of separation of concerns, provides the theoretical foundations of software architecture.

What are some typical software architectures?

Garlan and Shaw [1994] identify the following common architectural styles:

- pipes and filters
- objects
- implicit invocation
- layering
- repositories
- interpreters

It is beyond the scope of this text to describe these architectures, although Chapter 6 describes interpreters in some detail.

Other architectures include the more "mundane":

- distributed architectures such as client/server
- main program/subroutine organizations
- state-based architectures using FSMs
- interrupt driven reactive systems, such as those used in process-control (for example, the baggage inspection system)
- domain specific architectures, sometimes called "reference" architectures because they are typical of a certain application domain

Off-the-shelf or "precooked" architectures that have been or will be mentioned in this text include the Java-based J2EE architecture, Microsoft's. NET architecture, and the open source Struts (Model-View-Controller) architectures.

4.3 Software Design Modeling

What standard methodologies can be used for software design?

Two methodologies, process- or procedural-oriented and object-oriented design (OOD), are related to SA and OOA, respectively, and can be used to begin to perform the design activities from the SRS produced by either SA or structured design (SD). Each methodology seeks to arrive at a model containing small, detailed components.

What is procedural-oriented design?

Procedural-oriented design methodologies, such as SD, involve top-down or bottom-up approaches centered on procedural languages such as C and Fortran. The most common of these approaches utilizes design decomposition via Parnas partitioning.

Object-oriented languages provide a natural environment for information hiding, through encapsulation. The state, or data, and behavior, or methods, of objects are encapsulated and accessed only via a published interface or private methods. For example, in image processing systems, one may wish to define a class of type pixel, with characteristics (attributes) describing its position, color, brightness, and so on, and operations that can be applied to a pixel such as add, activate, and deactivate. The engineer may then wish to define objects of type image as a collection of pixels with other attributes. In some cases, expression of system functionality is easier to do in an object-oriented manner. We will discuss object-oriented design shortly.

What is SD?

SD is the companion methodology to SA. SD is a systematic approach concerned with the specification of the software architecture and involves a number of techniques, strategies, and tools. SD provides a step-by-step design process that is intended to improve software quality, reduce risk of failure, and increase reliability, flexibility, maintainability, and effectiveness.

What is the difference between SA and SD?

SA is related to SD in the same way that a requirements representation is related to the software design; that is, the former is functional and flat and the latter is modular and hierarchical. Data structure diagrams are then used to give information about logical relationships in complex data structures.

How do I go from SA to SD?

The transition mechanisms from SA to SD are manual and involve significant analysis and trade-offs of alternative approaches. Normally, SD proceeds from SA in the following manner. Once the context diagram is drawn, a set of DFDs is developed. The first DFD, the Level 0 diagram, shows the highest level of system abstraction. Decomposing processes to lower and lower levels until they are ready for detailed design renders new DFDs with successive levels of increasing detail. This decomposition process is called *leveling*.

In a typical DFD, boxes represent terminators that are labeled with a noun phrase that describes the system, agent, or device from which data enters or exits. Each process, depicted by a circle, is labeled as a verb phrase describing the operation to be performed on the data although it may be labeled with the name of a system or operation that manipulates the data. Solid arcs are used to connect terminators to processes and between processes to indicate the flow of data through the system. Each arc is labeled with a noun phrase that describes the data. Dashed arcs are discussed later. Parallel lines indicate data stores, which are labeled by a noun phrase naming the file, database, or repository where the system stores the data.

Each DFD should have between 3 and 9 processes only. The descriptions for the lowest level, or primitive, processes are called process specifications, or P-Specs, and are expressed in either structured English, pseudo-code, decision tables, or decision trees and are used to describe the logic and policy of the program (Figure 4.5).

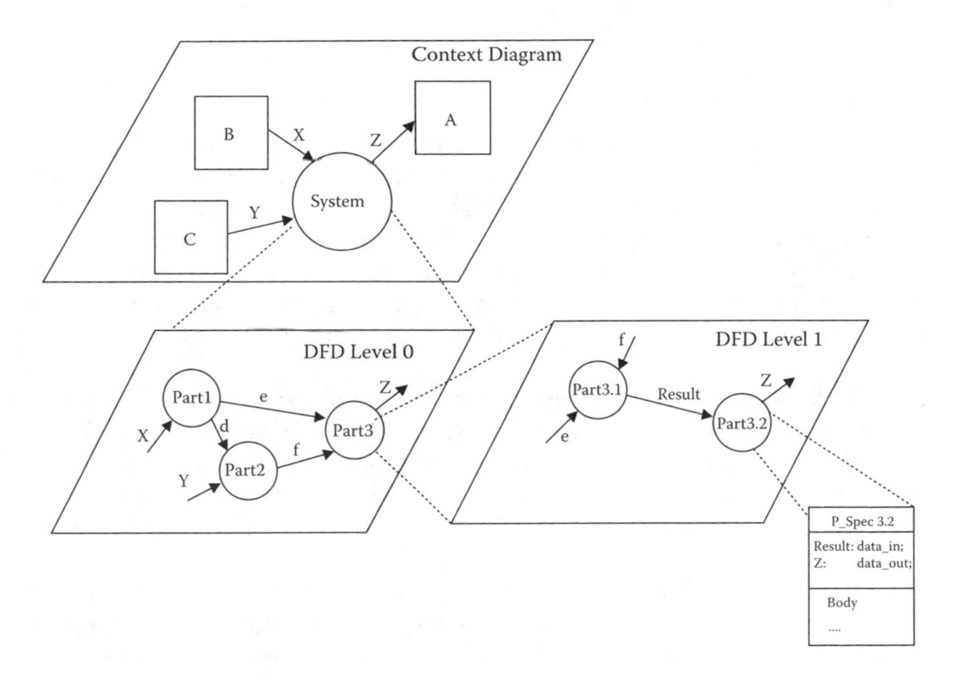

FIGURE 4.5
Context diagram evolution from context diagram to Level 0 DFD to Level 1 DFD and, finally, to a P-Spec, which is suitable for coding.

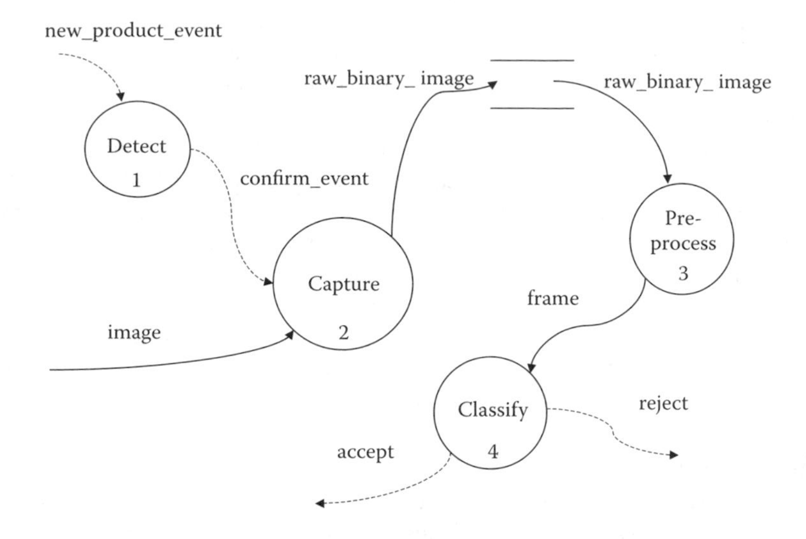

FIGURE 4.6
The Level 0 DFD for the baggage inspection system. The dashed arcs represent control flows, which are described later.

Can you give an example?

Returning to the baggage inspection system example, Figure 4.6 shows the Level 0 DFD. Here the details of the system are given at a high level. First, the system reacts to the arrival of a new product by confirming that the image data is available. Next, the system captures the image by buffering the raw data from the capture device to a file. Preprocessing of the raw data is performed to produce an image frame to be used for classification and generation of the appropriate control signals to the conveyor system.

Proceeding to the next level provides more detail for Processes 1, 2, 3, and 4. Process 1 is essentially an interrupt service routine assigned to a photodiode detector that senses when a new product for inspection reaches the designated point on the conveyer. Process 2 is a buffering routine whose characteristics depend on the specifications of the camera. Hence, without knowing these details, it is not possible to delve deeper into the design.

Figure 4.7 depicts the Level 1 DFD for Process 3. Notice how the internal Processes 3.1 and 3.2 are labeled to denote that they are a finer degree of detail of Process 3 shown in the Level 0 diagram. Successive levels of detail will follow a similar numbering system (e.g., 3.1.1, 3.1.2). This convention provides simple traceability from specification through design and on to the code. Continuing with the design example, Figure 4.8 shows the Level 1 DFD for Process 4.

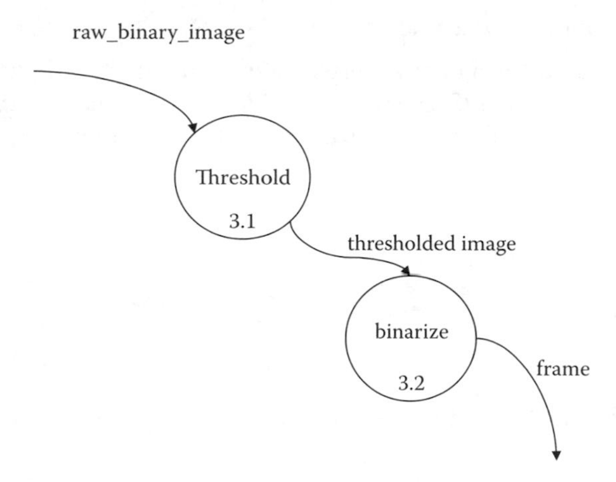

FIGURE 4.7
The Level 1 DFD for Process 3, pre-processing, of the baggage inspection system.

What is a data dictionary?

A data dictionary is a repository of data about data that describes every data entity in the system. The data dictionary is an essential component of the SD. The data dictionary includes entries for data flows, control flows, data stores, data elements, and control elements. Each entry is identified by name, range, rate, and units. The dictionary is organized alphabetically for ease of use.

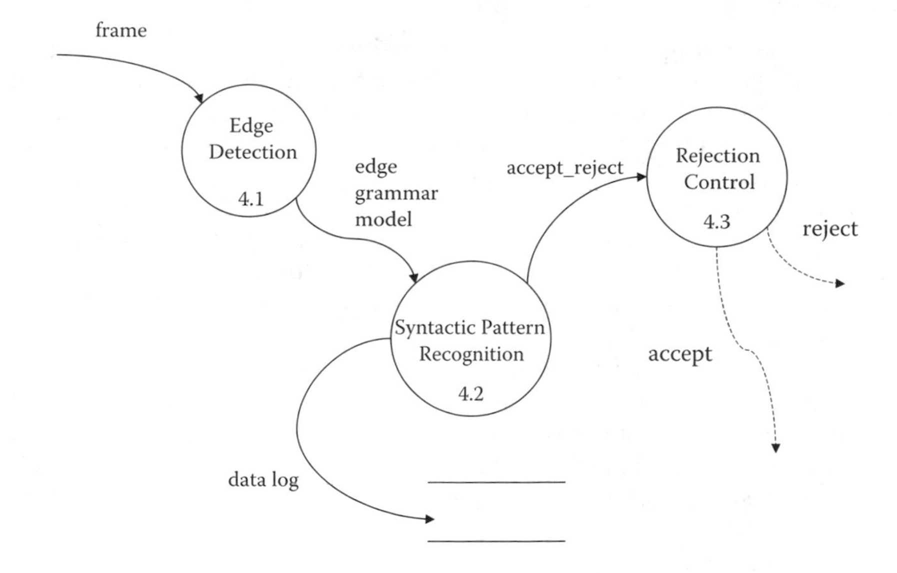

FIGURE 4.8
The Level 1 DFD for Process 4, classification, of the baggage inspections system.

What does a data dictionary look like?

There is no standard format, but every design element should have an entry in the data dictionary. Most CASE tools provide the data dictionary feature. For example, each entry could be organized to contain the following information:

Type
Name
Alias
Description
Found in

In particular, for the baggage inspection system, one entry might look like this:

Type: Data flow
Name: Binarized image
Alias: Image
Description: The raw binary image after it has been subjected to thresholding …
Found in:

The missing information for modules "Found in" will be added as the code is developed. In this way, data dictionaries help to provide substantial traceability between code elements.

Are there any problems with SDs?

There are several apparent problems in using SD to model the baggage inspection system, including difficulty in modeling time and events. For example, what if the baggage inspection system captures a new image in parallel with preprocessing of the last image capture?* Concurrency is not easily depicted in this form of SASD.

Another problem arises in the context diagram. Control flows are not easily translated directly into code, such as "reject" and "accept," because they are hardware dependent. In addition, the control flow does not really make sense because there is no connectivity between portions of it, a condition known as "floating."

Details of the detector and camera hardware also need to be known for further modeling of Process 1. What happens if the hardware changes? What if a different strategy for classification in Process 2 is needed? In the case of Process 3 (preprocessing), what if the algorithm or even the sensitivity levels change because of the installation of new hardware? In each case the changes

* This scenario would be desirable if the reject mechanism were further down the inspection line and the conveyor system were running at a high rate.

would need to propagate into the Level 1 DFD for each process, any subsequent levels, and, ultimately, the code.

Clearly making and tracking any of these changes is fraught with danger. Moreover, any change means that significant amounts of code would need to be rewritten, recompiled, and properly linked with the unchanged code to make the system work. None of these problems arise using the object-oriented paradigm.

How does SASD deal with timing?

In SASD, arcs made of dashed lines indicate the flow of control information and solid bars indicate "stored" control commands (control stores), which are left unlabeled.

Additional tools, such as Mealy FSMs, are used to represent the encapsulated behavior and process activations. The addition of the new control flows and control stores allows for the creation of a diagram containing only these elements. This is called a control flow diagram (CFD). These CFDs can be decomposed into control specifications (C-Specs), which can then be described by an FSM. The relationship between the control and process models is shown in Figure 4.9.

Can I use FSMs to derive a design?

Yes. One of the advantages of using FSMs in the SRS and later in the software design is that they are easily converted to code and test cases.

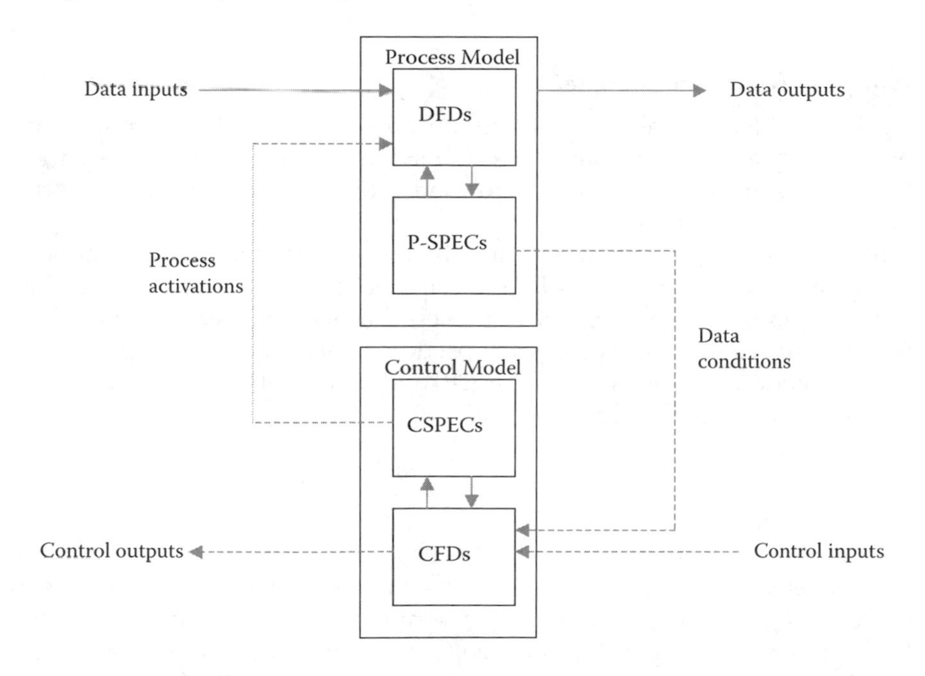

FIGURE 4.9
The relationship between the control and process model.

```
typedef states: (state1,...,staten);        {n is# of states}
        alphabet: (input1,...,inputn);
        table_row: array [1..n] of states;
procedure move_forward;        {advances FSM one state}
var
        state: states;
        input: alphabet;
        table: array [1..m] of table_row;  {m is the size of the alphabet}
begin
        repeat
          get(input); {read one token from input stream}
          state:=table[ord(input)] [state];    {next state}
          execute_process (state);
          until input = EOF;
end;
```

FIGURE 4.10
Pseudo-code that can implement the system behavior of the baggage inspection system depicted in Table 3.2.

Can you give me an example?

Again consider the baggage inspection system. The tabular representation of the state transition function (Table 3.2), which describes the system's high-level behavior, can be easily transformed into a design using the pseudo-code shown in Figure 4.10.

Each procedure associated with the three operational modes (operational, diagnostic, and calibration) will be structured code that can be viewed as executing in one of any number of process states at an instant in time. This functionality can be described by the pseudo-code shown in Figure 4.11.

The pseudo-code shown in Figure 4.10 and Figure 4.11 can be coded easily in any procedural language.

What is OOD?

OOD is an approach to systems design that views the system components as objects and data processes, control processes, and data stores that are encapsulated within objects. Early forays into OOD were led by attempts to reuse some of the better features of conventional methodologies, such as the DFDs and entity relationship models by reinterpreting them in the context of object-oriented languages. This approach can be seen in the UML. Over the

```
Procedure execute_process (state: states);
begin
        case state of
        state 1: process 1;           {execute process 1}
        state 2: process 2;           {execute process 2}

...
        staten: processn;             {execute process n}
end
```

FIGURE 4.11
FSM code for executing a single operational process of the baggage inspection system. Each process can exist in multiple states, allowing for partitioning of the code into appropriate modules.

last several years, the object-oriented framework has gained significant acceptance in the software engineering community.

What are the benefits of object-orientation?

Previous discussions highlighted some considerations concerning the appropriateness of the object-oriented paradigm to various application areas. The real advantages of applying object-oriented paradigms are the future extensibility and reuse that can be attained, and the relative ease of future changes.

Software systems are subject to near-continuous change: requirements change, merge, emerge, and mutate; target languages, platforms, and architectures change; and, most significantly, employment of the software in practice changes. This flexibility places considerable burden on the software design: how can systems that must support such widespread change be built without compromising quality? There are several basic rules of OOD that help us achieve these benefits.

What are the basic rules of OOD?

The following list of rules is widely accepted. All of these rules, except where noted, are from Martin [2002].

The Open/Closed Principle: Software entities (classes, modules, etc) should be open for extension, but closed for modification [Meyer 2000].

The Liskov Substitution Principle: Derived classes must be usable through the base class interface without the need for the user to know the difference [Liskov and Wing 1994].

The Dependency Inversion Principle: Details should depend upon abstractions. Abstractions should not depend upon details.

The Interface Segregation Principle: Many client specific interfaces are better than one general purpose interface.

The Reuse/Release Equivalency Principle: The granule of reuse is the same as the granule of release. Only components that are released through a tracking system can be effectively reused.

The Common Closure Principle: Classes that change together, belong together.

The Common Reuse Principle: Classes that aren't reused together should not be grouped together.

The Acyclic Dependencies Principle: The dependency structure for released components must be a directed acyclic graph. There can be no cycles.

The Stable Dependencies Principle: Dependencies between released categories must run in the direction of stability. The dependee must be more stable than the depender.

The Stable Abstractions Principle: The more stable a class category is, the more it must consist of abstract classes. A completely stable category should consist of nothing but abstract classes.

Once and only once (OAOO): Any aspect of a software system — be it an algorithm, a set of constants, documentation, or logic — should exist in only one place [Beck 1999].

A detailed discussion of these can be found in the aforementioned references.

What is the UML?

The UML is widely accepted as the *de facto* standard language for the specification and design of software-intensive systems using an object-oriented approach. By bringing together the "best-of-breed" in specification techniques, the UML has become a family of languages (diagram types). Users can choose which members of the family are suitable for their domain.

How does the UML help us with software design?

The UML is a graphical language based upon the premise that any system can be composed of communities of interacting entities and that various aspects of those entities, and their communication, can be described using the set of nine diagrams: use case, sequence, collaboration, statechart, activity, class, object, component, and deployment. Of these, five render behavioral views (use case, sequence, collaboration, statechart, and activity) while the remaining are concerned with architectural or static aspects.

With respect to embedded systems, these behavioral models are of interest. The use case diagrams document the dialog between external actors and the

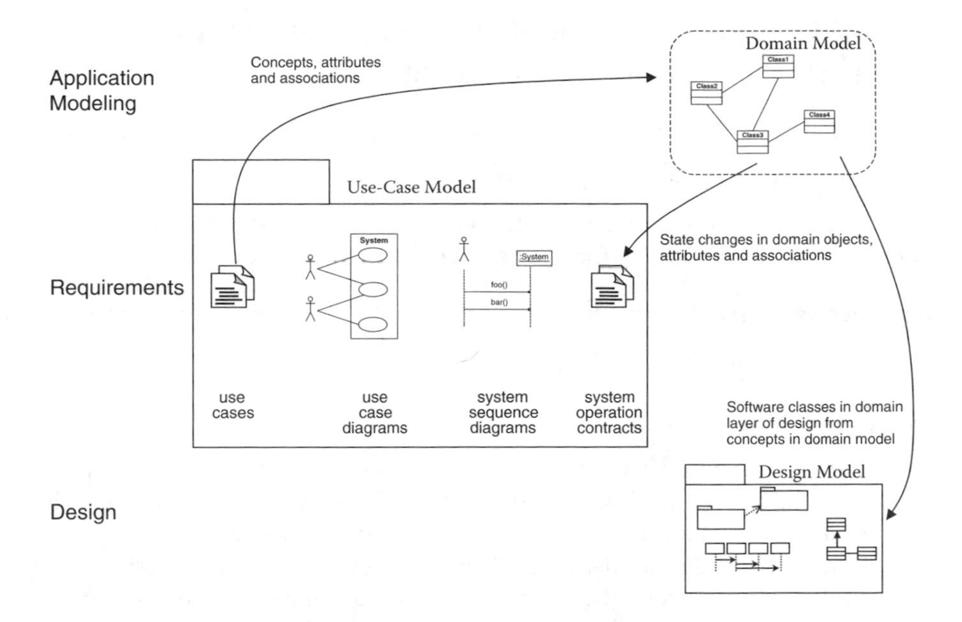

FIGURE 4.12
The UML and its role in specification and design. (Adapted from Larman, C., *Applying UML and Patterns,* Prentice-Hall, Englewood Cliffs, NJ, 2002.)

system under development. Sequence and collaboration diagrams describe interactions between objects. Activity diagrams illustrate the flow of control between objects. Statecharts represent the internal dynamics of active objects. The principle artifacts generated when using the UML and their relationships are shown in Figure 4.12.

While not aimed specifically at embedded system design, some notion of time has been included in the UML through the use of sequence diagrams.

What is the UML 2.0?

UML 2.0 is a revision to the UML that incorporates several improvements. At this writing, UML 2.0 is only beginning to replace UML as the *de facto* standard. These improvements include:

- New base classes that provide the foundation for UML modeling constructs.
- Object constraint language, a formal method that can be used to better describe object interactions.
- An improved diagram meta-model that allows users to model systems from four viewpoints:
  - Static models (e.g., class diagrams)
  - Interaction (e.g., using sequence diagrams)

- Activity (i.e., to describe the flow of activities within a system)
- State (i.e., to create FSMs using statecharts)

These viewpoints are intended to be complementary [France et al. 2006].

Can you give an example of an OOD?

Appendix B contains an SDS for the wet well control system. Appendix C contains the object models for the wet well control system.

4.4 Pattern-Based Design

What is a pattern?

Informally, a pattern is a named problem-solution pair that can be applied in new contexts, with advice on how to apply it in those situations. The formal definition of a pattern is not consistent in the literature.

Patterns can be distinguished as three types: architectural, design, and idioms. An architectural pattern occurs across subsystems, a design pattern occurs within a subsystem but is independent of the language, and an idiom is a low-level pattern that is language specific.

What is the history of patterns?

Christopher Alexander first introduced the concept of design patterns for architecture and town planning. He realized that the same problems were encountered in the design of buildings and once an elegant solution was found it could be applied repeatedly. "Each pattern describes a problem which occurs over and over again in our environment, and then describes the core of the solution to that problem, in such a way that you can use this solution a million times over, without ever doing it the same way twice" [Alexander et al. 1977].

Patterns were first applied to software in the 1980s by Ward Cunningham, Kent Beck, and Jim Coplien. Then the famous "Gang of Four" book, *Design Patterns: Elements of Reusable Object-Oriented Software* [Gamma et al. 1995] popularized the use of patterns.

Why do we need patterns?

Developing software is hard and developing reusable software is even harder. Designs should be specific to the current problem, but general enough to address future problems and requirements. Experienced designers know not to solve every problem from first principles, but to reuse solutions encountered previously. They find recurring patterns and then use them as a basis for new designs. This is simply an embodiment of the Principle of Generality.

What are the benefits of patterns?

First, design patterns help in finding appropriate objects, in determining object granularity, and in designing in anticipation of change. At the design level, design patterns enable large-scale reuse of software architectures by capturing expert knowledge and making this expertise more widely available. Finally, patterns help improve developer vocabulary (I am always amazed to listen to a group of software engineers with substantial pattern knowledge discuss designs in what seems like a different language). While the terminology might be unfamiliar to those not knowing the particular pattern language, the efficiency of information exchange is very high. Some people also contend that learning patterns can help in learning object-oriented technology.

What do patterns look like?

In general, a pattern has four essential elements:

a name

a problem description

a solution to the problem

the consequences of the solution

The "name" is simply a convenient handle for the pattern itself. Some of the names can be rather humorous (such as "flyweight" and "singleton"), but are intended to evoke an image to help remind the user of the intent of the pattern.

The "problem" part of the pattern template states when to apply the pattern; that is, it explains the problem and its context. The problem statement may also describe specific design problems such as how to represent algorithms as objects. The problem statement may also describe class structures that are symptomatic of an inflexible design and possibly include conditions that must be met before it makes sense to apply the pattern.

The "solution" describes the elements that make up the design, although it does not describe a particular concrete design or implementation. Rather, the solution provides how a general arrangement of objects and classes solves the problem.

Finally, "consequences" show the results and trade-offs of applying pattern. It might include the impact of the pattern on space and time, language and implementation issues, and flexibility, extensibility, and portability. The consequences are critical for evaluating alternatives.

What are the "GRASP" patterns?

The GRASP (general principles in assigning responsibilities) patterns are a fairly high-level set of patterns for design set forth by Craig Larman [2002]. The GRASP patterns are:

- Creator
- Controller
- Expert

- Low coupling
- High cohesion
- Polymorphism
- Pure fabrication
- Indirected
- Protected variations

Let us look at each of these patterns briefly, giving only name, problem, and solution in an abbreviated pattern template. The consequence of using each of these, generally, is a vast improvement in the OOD.

Name: Creator

Problem: Who should be responsible for creating a new instance of some class?

Solution: Assign Class B the responsibility to create an instance of class A if one or more of the following is true:

- B aggregates A objects.
- B contains A objects.
- B records instances of A objects.
- B closely uses A objects.

B has the initializing data that will be passed to A when it is created.

For example, in the baggage inspection system, the class "camera" would create objects from the class `baggage_image`.

Name: Controller

Problem: Who should be responsible for handling an input system event?

Solution: Assign the responsibility for receiving or handling a system event message to a class that represents the overall system or a single use case scenario for that system.

In the baggage inspection system, the "baggage_inspection_system" class would be responsible for `new_baggage` events.

Name: Expert

Problem: What is a general principle of assigning responsibilities to objects?

Solution: Assign a responsibility to the information expert — the class that has the information necessary to fulfill the responsibility.

For example, in the baggage inspection system the class `threat_detector` would be responsible for identifying objects of the class baggage as being a possible threat.

Name: Low Coupling

Problem: How do we support low dependency, low change impact, and increased reuse?

Solution: Assign the responsibility so the coupling remains low.

Applying separation of concerns or Parnas partitioning principles would be helpful here.

Name: Polymorphism

Problem: How can we design for varying, similar cases?

Solution: Assign a polymorphic operation to the family of classes for which the cases vary.

For example, in the baggage inspection system we probably have different baggage object types (for example, suitcase, golf club case, baby seat, etc.) and the algorithm for inspecting the images derived from each of these should be different. But the method that scans each image should be determined at run-time depending upon the object, not through the use of a case statement.

Name: Pure fabrication

Problem: Where can we assign responsibility when the usual options based on Expert lead to high coupling and low cohesion?

Solution: Create a "behavior" class whose name is not necessarily inspired by the domain vocabulary in order to reduce coupling and increase cohesion.

For example, in the baggage inspection system we might contrive a hobo class to describe a particular type of luggage that is nonconventional (for example, a cardboard box, a laundry bag, or other unusual container).

Name: Indirection

Problem: How do we reduce coupling?

Solution: Assign a responsibility to an intermediate object to decouple two components.

Here we might create a `second_look` class to deal with baggage that might need re-imaging.

Name: Protected Variations

Problem: How can we design components so that the variability in these elements does not have an undesirable impact on other elements?

Solution: Identify points of likely variation or instability; assign responsibilities to create a stable interface around them.

TABLE 4.1

The Set of Design Patterns Popularized by the"Gang of Four"

Creational	Behavioral	Structural
Abstract factory	Chain of responsibility	Adapter
Builder	Command	Bridge
Factory method	Interpreter	Composite
Prototype	Iterator	Decorator
Singleton	Mediator	Facade
	Memento	Flyweight
	Observer	Proxy
	State	
	Strategy	
	Template method	
	Visitor	

Source: Gamma, E., Helm, R., Johnson, R., and Vlissides, J., *Design Patterns: Elements of Reusable Object-Oriented Software,* Addison-Wesley, Boston, MA, 1995.

This principle is essentially the same as information hiding and the open-closed principle.

What are the Gang of Four patterns?

This set of design patterns was first introduced by Gamma, Helm, Johnson, and Vlissides (the "Gang of Four" or "GoF") and popularized in a well-known text [Gamma et al. 1995]. They describe 23 patterns organized by creational, behavioral, or structural intentions (Table 4.1).

Many of the GoF patterns perfectly illustrate the convenience of the name in suggesting the application for the pattern. For example, the flyweight pattern provides a design strategy when a large number of small-grained objects will be needed, such as baggage objects in the baggage inspection system. The singleton pattern is used when there will be a single instance of an object, such as the single instance of a baggage inspection system object. In fact, the singleton pattern can be used as the base class for any system object.

Explication of each pattern is beyond the scope of this text. Table 4.1 is provided for illustration only. Interested readers are encouraged to consult Gamma et al [1995]. However, the three Gang of Four pattern types will be discussed in general terms.

What are creational patterns?

Creational patterns are connected with object creation and they allow for the creation of objects without actually knowing what you are creating beyond the interfaces. Programming in terms of interfaces embodies information hiding. Therefore, we try to write as much as possible in terms of interfaces.

What are structural patterns?

Structural patterns are concerned with organization classes so that they can be changed without changing the other code. Structural patterns are "static model" patterns; that is, they represent structure that should not change.

What are behavioral patterns?

Behavioral patterns are concerned with runtime (dynamic) behavior of the program. They help define the roles of objects and their interactions, but being dynamic they do not contain much, if any, structure

Are there any other pattern sets?

There are many other pattern sets. For example, there is a well-known set of architecture and design patterns [Buschmann et al. 1996], analysis patterns [Fowler 1996], and literally dozens of others.

What are the drawbacks of patterns?

Patterns do not lead to direct code reuse. Direct code reuse is the subject of software libraries. Rather, patterns lead to reusable design and architectures, which can be converted to code.

Patterns are deceptively simple. While it might be easy enough to master the names of the patterns and memorize their structure visually, it is not so easy to see how they can lead to design solutions. This benefit takes a great deal of education and experience.

Teams may suffer from pattern overload, meaning that the quest to use pattern-based techniques can be an obsession and an end itself rather than the means to an end. Patterns are not a silver bullet. Rather, they provide another approach to solving design problems.

Finally, integrating patterns into a software development process is a labor-intensive activity; therefore, immediate benefits from a patterns program may not be realized.

4.5 Design Documentation

Is there a standard format for SDS?

No. There are many different variations on the SDS (also called SDD). IEEE Standard 1016-1998, *IEEE Recommended Practice for Software Design Descriptions,* is one possible resource that you can consult. But, every company uses a different template for this documentation.

How do I achieve traceability from requirements through design and testing?

One way to achieve these links is through the use of an appropriate numbering system throughout the documentation. For example, a requirement numbered 3.2.2.1 would be linked to a design element with a similar number (the numbers don't have to be the same so long as the annotation in the document provides traceability). These linkages are depicted in Figure 4.13. Although the documents shown in the figure have not been introduced yet, the point to be made is that the documents are all connected through appropriate referencing and notation.

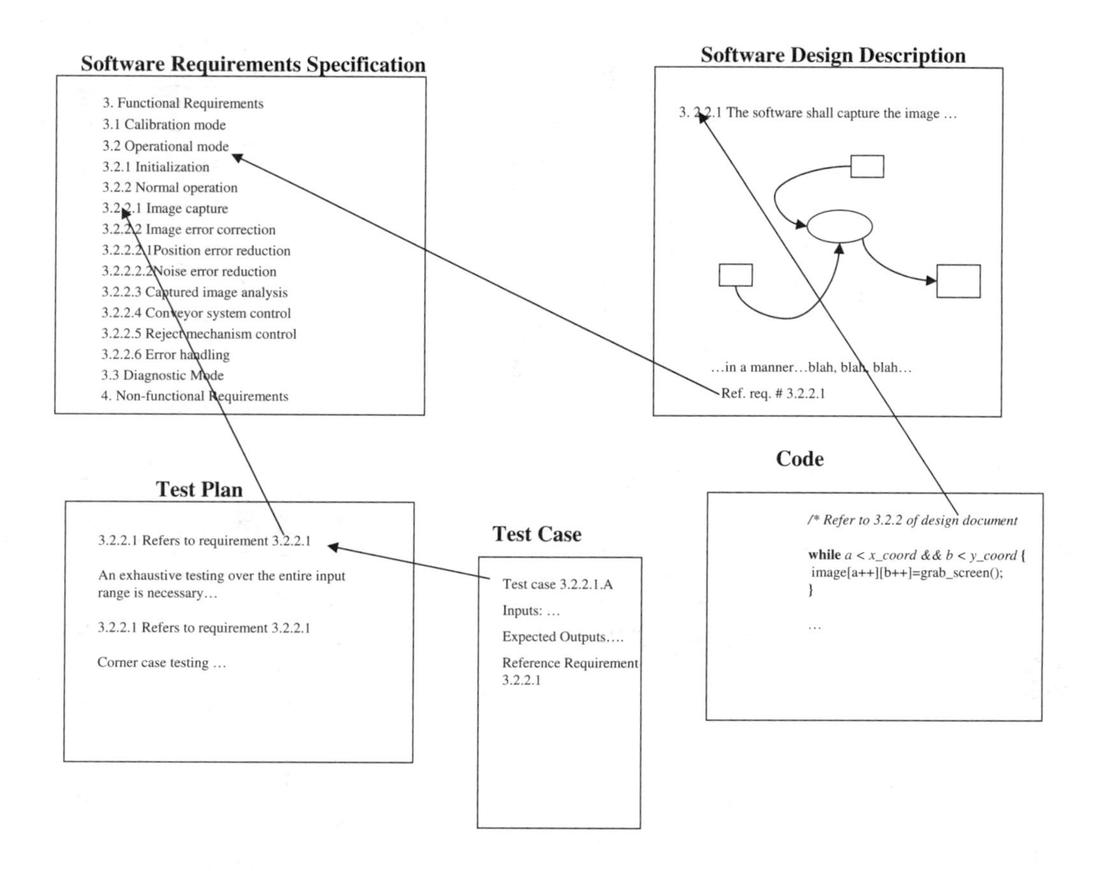

FIGURE 4.13
Linkages between software documentation and code. In this case, the links are achieved through similarities in numbering and specific reference to the related item in the appropriate document.

TABLE 4.2

A Traceability Matrix Corresponding to Figure 4.13 Sorted by Requirement Number

Requirement Number	SDD Reference Number	Test Plan Reference Number	Code Unit Name	Test Case Number
3.1.1.1	3.1.1.1 3.2.4	3.1.1.1 3.2.4.1 3.2.4.3	Simple_fun	3.1.1.A 3.1.1.B
3.1.1.2	3.1.1.2	3.1.1.2	Kalman_filter	3.1.1.A 3.1.1.B
3.1.1.3	3.1.1.3	3.1.1.3	Under_bar	3.1.1.A 3.1.1.B 3.1.1.C

Figure 4.13 is simply a graphical representation of the traceable links. In practice, a traceability matrix is constructed to help cross reference documentation and code elements (Table 4.2). The matrix is constructed by listing the relevant software documents and the code unit as columns, and then each software requirement as rows.

Constructing the matrix in a spreadsheet software package allows for providing multiple matrices sorted and cross-referenced by each column as needed. For example, a traceability matrix sorted by test case number would be an appropriate appendix to the text plan.

The traceability matrices are updated at each step in the software life cycle. For example, the column for the code unit names (e.g., procedure names, object class) would not be added until after the code is developed.

Finally, a way to foster traceability between code units is through the use of data dictionaries.

Can you give an example of a design document?

Appendix B contains an SDS for the wet well control system. Appendix C contains the object models for the wet well control system. The latter document is sometimes used as a supplement to an SDS.

4.6 Further Reading

Alexander, C., Ishikawa, S., and Silverstein, M, *A Pattern Language*, Oxford University Press, New York, 1977.

Beck, K., *Extreme Programming: Embracing Change*, Addison-Wesley, Boston, MA, 1999.

Buschmann, F., Meunier, R., Rohnert, H., Sommerlad, P., and Stal, M., *Pattern-Oriented Software Architecture, Volume 1: A System of Patterns*, John Wiley & Sons, New York, 1996.

Fowler, M., *Analysis Patterns: Reusable Object Models*, Addison-Wesley, Boston, MA, 1996.

France, R.B., Ghosh, S., Dinh-Trong, T., and Solberg, A., Model-driven development using UML 2.0: promises and pitfalls, *Computer*, 39(2), 59–66, 2006.

Gamma, E., Helm, R., Johnson, R., and Vlissides, J., *Design Patterns: Elements of Reusable Object-Oriented Software*, Addison-Wesley, Boston, MA, 1995.

Garlan, D. and Shaw, M., An Introduction to Software Architecture, Technical Report CMU/SEI-94-TR-21, January 1994, pp. 1–42.

Kruchten, P., Obbink, H., and Stafford, J., The past, present, and future of software architecture, *IEEE Software*, 23(2), 22–30, 2006.

Laplante, P.A., *Software Engineering for Image Processing Systems*, CRC Press, Boca Raton, FL, 2004.

Larman, C., *Applying UML and Patterns*, Prentice-Hall, Englewood Cliffs, NJ, 2002.

Levine, D. and Schmidt, D., *Introduction to Patterns and Frameworks*, Washington University, St Louis, 2000.

Liskov, B. and Wing, J., A behavioral notion of subtyping, *ACM Trans. Program. Lang. Syst.*, 16(6), 1811–1841, 1994.

Martin, R.C., *Agile Software Development: Principles, Patterns, and Practices*, Prentice-Hall, Englewood Cliffs, NJ, 2002.

Meyer, B., *Object-Oriented Software Construction*, 2nd ed., Prentice Hall, Englewood Cliffs, NJ, 2000.

Parnas, D. L. On the criteria to be used in decomposing systems into modules, *Commun. ACM*, 15(12) , 1053–1058, 1972.

Pressman, R.S. *Software Engineering: A Practitioner's Approach*, 6th ed., McGraw-Hill, New York, 2005.

Sommerville, I., *Software Engineering*, 7th ed., Addison-Wesley, Boston, MA, 2005.

5

Building Software

Outline

- Programming languages
- Software construction tools
- Becoming a better code developer

5.1 Introduction

When I talk of "building software," I mean using tools to translate the designed model of the software system into runnable behavior using a programming language. The tools involved include the means for writing the code (a text editor), the means for compiling that code (the programming language compiler or interpreter), debuggers, and tools for managing the build of software on local machines and between cooperating computers across networks, including the Web.

Other tools for building software include version control and configuration management software and the integrated environments that put them together with the language compiler to provide for seamless code production. With these tools, the build process can be simple, especially when the specification, design, testing, and integration of the software are well planned and executed.

In this chapter I will discuss the software engineering aspects of programming languages and the software tools used with them. Much of this discussion has been adapted from my book *Software Engineering for Image Processing Systems* [Laplante 2004a].

5.2 Programming Languages

Is the study of programming languages really of interest to the software engineer?

Yes. A programming language represents the nexus of design and structure. But a programming language is really just a tool and the best software developers are known for the quality of their tools and their skill with them. This skill can only be obtained through a deep understanding of the language itself and the peculiarities of a particular implementation of a language as seen in the compiler.

What happens when software behaves correctly, but is poorly written?

Misuse of the programming language is very often the cause in the reduction of the desirable properties of software (such as maintainability and readability) and the increase in undesirable properties (such as fragility and viscosity).

But I have been writing code since my first programming course in college. Surely, I don't need any lessons in programming languages, do I?

For many engineers, Fortran, BASIC, assembly language, C, or Pascal were the first programming languages they learned. This is an important observation because many engineers learned how to program, but not how to program well. Bad habits are hard to break. Moreover, these languages are not object-oriented, and many new systems are now being implemented using the object-oriented paradigm. So, when today's engineer is asked to implement a new system in C++ or Java, he may not be using proper methodology and, therefore, losing the benefits of modern programming practices even though he can get the system to work.

What about working with legacy code?

A great deal of legacy code is written in Fortran, COBOL, assembly language, Ada, Jovial, C, or any number of other exotic and antiquated languages. It is always hard to perform maintenance work on an old system implemented in one of these languages.

So how many programming languages are there?

There are literally hundreds of them, many of which are arcane or so highly specialized that there is no real benefit to discussing them. In the following, I discuss a few of the more frequently encountered ones. First I provide a framework for that discussion by describing the landscape of programming languages.

5.2.1 Programming Language Landscape

What does the programming landscape look like?

There are two different ways to partition the world of programming languages:

imperative, functional, logic, and other
procedural or declarative

Imperative and procedural languages, which refer to the same set of languages, are the most commonly found in the industry.

What are imperative, functional, and logic languages?

One way to partition programming languages is by considering whether they are imperative, functional, logic, or other. Imperative languages involve assignment of variables (the most widely used languages are imperative, for example, C, Java, Ada95, Fortran, and Visual Basic). Functional (applicative) languages employ repeated application of some function (for example, LISP, Scheme, and Haskell). Logic programming involves a formal statement of the problem without any specification of how it is solved (for instance, PROLOG). "Other" encompasses every other language that does not neatly fit into one of the other categorizations.

What are procedural and declarative languages?

The other way to partition programming languages is by considering whether they are procedural or declarative. Procedural languages are defined by a series of operations and include all of the languages named as imperative. "Procedural" also describes a style of programming that is not object-oriented. Declarative (nonprocedural) languages involve specification of rules defining the solution to a problem (for example, PROLOG, Spreadsheets).

Then what are object-oriented languages?

Object-oriented programming languages are those characterized by data abstraction, inheritance, and polymorphism. Data abstraction has been previously defined. Inheritance allows the software engineer to define new objects in terms of previously defined objects so that the new objects "inherit" properties. Function polymorphism allows the programmer to define operations that behave differently depending on the type of object involved. For example, a filter operation would act differently depending on the type of image and filtering needed. How the filter operation is applied is implemented at run time.

5.2.2 Programming Features and Evaluation

Which programming language is better?

Unfortunately, the answer is, "it depends." It is like the question "which is better, Italian food or Chinese food?" You can't make objective comparisons because it depends upon whom you ask. So to it is with programming languages.

Now, some programming languages are better for certain applications or situations; for example, C is better for certain embedded systems, scripting languages are better for rapid prototyping, and so forth. Indeed, each programming language offers its own strengths and weaknesses with respect to specific applications domains.

So what is the best way to evaluate a programming language?

I like Cardelli's [1996] evaluation criteria. He classifies languages along the following dimensions.

- **Economy of execution** — How fast does the program run?
- **Economy of compilation** — How long does it take to go from sources to executables?
- **Economy of small-scale development** — How hard must an individual programmer work?
- **Economy of large-scale development** — How hard must a team of programmers work?
- **Economy of language features** — How hard is it to learn or use a programming language?

But as discussions around these dimensions can get somewhat theoretical, I will use more tangible "visible" features to discuss the more commonly encountered programming languages.

What do you mean by visible features of programming languages?

There are several programming language features that stand out, particularly in procedural languages, which are desirable for use in the kind of software systems with which engineers deal. These are as follows:

- versatile parameter-passing mechanisms
- dynamic memory allocation facilities
- strong typing
- abstract data typing
- exception handling
- modularity

These language features help promote the desirable properties of software and best engineering practices.

Why should I care about parameter-passing techniques?

Use of parameter lists is likely to promote modular design because the interfaces between the modules are clearly defined. There are several methods of parameter passing, but the three most commonly encountered are call-by-value, call-by-reference, and global variables.*

Clearly defined interfaces can reduce the potential of untraceable corruption of data by procedures using global access. However, both call-by-value and call-by-reference parameter-passing techniques can impact performance when the lists are long because interrupts are frequently disabled during parameter passing to preserve the time correlation of the data passed. Moreover, call-by-reference can introduce subtle function side effects that depend on the compiler.

What is call-by-reference?

In call-by-reference or call-by-address, the address of the parameter is passed by the calling routine to the called procedure so that it can be altered there. Execution of a procedure using call-by-reference can take longer than one using call-by-value because indirect mode instructions are needed for any calculations involving the variables passed in call-by-reference. However, in the case of passing large data structures between procedures, it is more desirable to use call-by-reference because passing a pointer to a large data structure is more efficient than passing the structure field by field.

What is call-by-value?

In call-by-value parameter passing, the value of the actual parameter in the subroutine or function call is copied into the formal parameter of the procedure. Because the procedure manipulates the formal parameter, the actual parameter is not altered. This technique is useful when either a test is being performed or the output is a function of the input parameters. For example, in an edge detection algorithm, an image is passed to the procedure and some description of the location of the edges is returned, but the image itself need not be changed.

When parameters are passed using call-by-value, they are copied onto a run-time stack at considerable execution time cost. For example, large data structures must be passed field by field.

What about global variables?

Global variables are variables that are within the scope of all modules of the software system. This usually means that references to these variables can be made in direct mode and thus are faster than references to variables passed via parameter lists. For example, in many image-processing applications,

* There are three other historical parameter-passing mechanisms: call-by-constant, which was removed almost immediately from the Pascal language; call-by-value-result, which is used in Ada; and call-by-name, which was a mechanism peculiar to Algol-60.

global arrays are defined to represent images. Hence, costly parameter passing can be avoided.

Global variables can be dangerous because reference to them may be made by unauthorized code, thus introducing subtle bugs. For this and other reasons, unwarranted use of global variables is to be avoided. Global parameter passing is recommended only when timing warrants and its use must be clearly documented.

How do I choose which parameter-passing technique to use?

The decision to use one method of parameter passing or the other represents a trade-off between good software engineering practice and performance needs. For example, often timing constraints force the use of global parameter passing in instances when parameter lists would have been preferred for clarity and maintainability.

What is recursion?

Most programming languages provide for recursion in that a procedure can call itself. For example, in pseudo-code:

```
foobar( int x, int y)
{
 if
 (x < y) foobar( y-x, x);
 else
   return (x);
}
```

Here, foobar is a recursive procedure involving two integers. Invoking foobar(1,2) will return the value 1.

Recursion is widely used in many mathematical algorithms that underlie engineering applications. Recursion can simplify programming of non-engineering applications as well.

Are there any drawbacks to recursive algorithm formulations?

Yes. While recursion is elegant and often necessary, its adverse impact on performance must be considered. Procedure calls require the allocation of storage on one or more stacks for the passing of parameters and for storage of local variables.

The execution time needed for the allocation and deallocation, and for the storage of those parameters and local variables can be costly. In addition, recursion necessitates the use of a large number of expensive memory and register indirect instructions.

Finally, the use of recursion often makes it impossible to determine the size of run-time memory requirements. Thus, iterative techniques such as

while and **for** loops must be used if performance prediction is crucial or in those languages that do not support recursion.

What does "dynamic memory allocation" mean?

This means that memory storage requirements need to be known before the program is written so that memory can be allocated and deallocated as needed. The capability to dynamically allocate memory is important in the construction and maintenance of many data structures needed in a complex engineering system. While dynamic allocation can be time consuming, it is usually necessary, especially when creating intermediate data structures needed in many engineering algorithms.

What are some examples of dynamic allocation use?

Linked lists, trees, heaps, and other dynamic data structures can benefit from the clarity and economy introduced by dynamic allocation. Furthermore, in cases where just a pointer is used to pass a data structure, then the overhead for dynamic allocation can be quite reasonable. When writing such code, however, care should be taken to ensure that the compiler will pass pointers to large data structures and not the data structure itself.

What is meant by "strong typing" in a programming language?

Strongly typed languages require that each variable and constant be of a specific type (e.g., "float," "Boolean," or "integer") and that each be declared as such before use. Generally, high-level languages provide integer and floating point types, along with Boolean, character, and string types. In some cases, abstract data types are supported. These allow programmers to define their own type along with the associated operations.

Strongly typed languages prohibit the mixing of different variable types in operations and assignments, and thus force the programmer to be precise about the way data are to be handled. Precise typing can prevent corruption of data through unwanted or unnecessary type conversion.

Weakly typed languages either do not require explicit variable type declaration before use (those of you familiar with old versions of Fortran or BASIC are recognize this concept) or do not prohibit mixing of types in arithmetic operations. Because these languages generally perform mixed calculations using the type that has the highest storage complexity, they must promote all variables to that type. For example, in C, the following code fragment illustrates automatic promotion and demotion of variable types:

```
int x,y;
float k,l,m;
   .
   .
j = x*k+m;
```

Here the variable x will be promoted to a float (real) type and then multiplication and addition will take place in floating point. Afterward, the result will be truncated and stored in j. The performance impact is that hidden promotion and more time-consuming arithmetic instructions can be generated with no additional accuracy. In addition, accuracy can be lost due to the truncation or, worse, an integer overflow can occur if the floating-point value is larger than the allowable integer value.

Programs written in languages that are weakly typed need to be scrutinized for such effects. Some C compilers will catch type mismatches in function parameters. This can prevent unwanted type conversions.

What is exception handling?

Certain languages provide facilities for dealing with errors or other anomalous conditions that arise during program execution. These conditions include the obvious, such as floating-point overflow, square root of a negative, divide-by-zero, and image-related conditions such as boundary violation, wraparound, and pixel overflow.

The capability to define and handle exceptional conditions in high-level languages aids in the construction of interrupt handlers and other code used for real-time event processing. Moreover, poor handling of exceptions can degrade performance. For example, floating-point overflow errors can propagate bad data through an algorithm and instigate time-consuming error recovery routines.

Which languages have the best exception handling facilities?

Java has excellent exception handling through its "try, catch, finally" approach, which is used in many mainstream object-oriented languages. In Java, the structure looks something like this:

```
try
{
     // do something
}
catch (Exception1)
{
     // what to do if something goes wrong
}
catch (Exception2)
{
     // what to do if something else goes wrong
}
...
```

```
finally
{
    // what to do if all else fails
}
```

In this way, foreseeable "risky" computation can be handled appropriately, rather than relying on the operating system to use its standard error handling, which may be inadequate.

Ada95, C++, and C# also have excellent exception handling capabilities. ANSI-C provides some exception handling capability through the use of signals.

What is meant by modularity?

Procedural languages that are amenable to the principle of information hiding and separation of concerns tend to make it easy to construct subprograms. While C and Fortran both have mechanisms for this (procedures and subroutines), other languages such as Ada95 (which can be considered either procedural or object-oriented) tend to foster more modular design because of the requirement to have clearly defined input and outputs in the module parameter lists.

In Ada, the notion of a package exquisitely embodies the concept of Parnas information hiding. The Ada package consists of a specification and declarations that include its public or visible interface and its invisible or private parts. In addition the package body, which has further externally invisible components, contains the working code of the package. Packages are separately compliable entities, which further enhances their application as black boxes.

In Fortran, there is the notion of a subroutine and separate compilation of source files. These language features can be used to achieve modularity and design abstract data types.

The C language also provides for separately compiled modules and other features that promote a rigorous top-down design approach, which should lead to a good modular design.

While modular software is desirable, there is a price to pay in the overhead associated with procedure calls and parameter passing. This adverse effect should be considered when sizing modules.

Do object-oriented languages support a form of modularity?

Object-oriented languages provide a natural environment for information hiding. For example, in image processing systems, it might be useful to define a class of type pixel, with attributes describing its position, color, and brightness; and operations that can be applied to a pixel such as add, activate, deactivate, and so on. It might also be desirable to define objects of type image as a collection of pixels with other attributes of width, height, and so on.

In some cases, expression of system functionality is easier to do in an object-oriented manner.

What is the benefit of object-orientation from a programming perspective?

Object-oriented techniques can increase programmer efficiency, reliability, and the potential for reuse. More can be said on this subject, but the reader is referred to the references at the end of the chapter [Gamma 1995, Meyer 2000].

5.2.3 Brief Survey of Languages

Can you apply the micro properties just discussed to some of the more commonly used programming languages?

For purposes of illustrating the aforementioned language properties, let us review some of the more widely used languages in engineering systems. The languages are presented in alphabetical order:

- Ada95
- assembly language
- C
- C++
- Fortran
- Java

Functional languages, such as LISP and ML, have been omitted from the discussions. This is simply because their use in this setting is rare.

Can you tell me about Ada95?

First introduced in 1983, Ada83 was originally intended to be the mandatory language for all U.S. DoD projects, which included many embedded real-time systems. Ada83, had significant problems, though, and Ada95 was introduced to deal with these problems. Throughout the text when the term "Ada" is used, "Ada95" is meant.

Ada95 was the first internationally standardized object-oriented programming language, athough it is both procedural and object-oriented depending upon how the programmer uses it.

Ada95 includes extensions to help deal with scheduling, resource contention, and synchronization. Features such as tagged types, packages, and protected units helped make Ada95 an object-oriented language.

Is Ada still used?

Ada95 is used in some university curricula to teach programming. But Ada95 never lived up to its promise of universality because users found the language to be too bulky and inefficient. Nevertheless, the language is staging somewhat of a comeback, particularly because of the availability of open source compilers for Linux (Linux is an open source derivative of the Unix operating system). Ada95 can compile to Java byte code, giving it broader applicability. I have had several students in recent years maintaining legacy defense systems written in Ada. It is important to mention as well that several million lines of code for the International Space Station (embedded flight and ground support) are written in Ada95.

What about assembly language?

Though lacking most of the features discussed for the high-level languages, assembly language does have certain advantages in some embedded control systems in that it provides more direct control of the computer hardware. Assembly language is sometimes used in tandem with a high-order language, especially C. This approach can be used to optimize for a particular platform, particularly the Intel 80 x 86 family, which has grown considerably in terms of instructions and capability.

Unfortunately, because of its unstructured and limited abstraction properties, and because it varies widely from machine to machine, coding in assembly language is usually difficult to learn, tedious, and error-prone. The resulting code is also nonportable.

How do programmers use assembly language today?

Assembly code is used in rare cases where the compiler does not support certain macro-instructions, or when the timing constraints are so tight that hand tuning is needed to produce optimal code. Even the most time-critical system will likely have 90% of the code written in a high-order language, while the rest is written in assembly language.

When should assembly language be used?

If ever, assembly language programming should be limited to use in very tight timing situations or in controlling hardware features that are not supported by the compiler. The use of assembly language needs to be well-documented.

C is my favorite programming language. When can it be used?

The C programming language, invented around 1971, is a good language for "low-level" programming. This is because C is descended from the language BCPL (whose successor, C's parent, was "B"), which supported only one type — machine word. Consequently, C supported machine-related objects like characters, bytes, bits, and addresses, which could be handled directly in high-level language. These entities can be manipulated to control interrupt controllers, CPU registers, and other hardware needed in embedded systems. C is sometimes used as a high-level cross-platform assembly language.

What other useful features does C have?

C provides special variable types such as register, volatile, static, and constant, which allow for control of code generation at the high-order language level. For example, declaring a variable as a register type indicates that it will be used frequently. This encourages the compiler to place such a declared variable in a register, which often results in smaller and faster programs. C supports call-by-value only, but call-by-reference can be implemented by passing the pointer to anything as a value.

Variables declared as type volatile are not optimized by the compiler. This is useful in handling memory-mapped Input/Output (I/O) and other instances where the code should not be optimized.

How does C handle exceptions?

The C language provides for exception handling through the use of signals, and two other mechanisms, `setjmp` and `longjmp`, are provided to allow a procedure to return quickly from a deep level of nesting. This is a useful feature in procedures requiring an abort. The `setjmp` procedures call, which is really a macro (but often implemented as a function), saves environment information that can be used by a subsequent `longjmp` library function call. The `longjmp` call restores the program to the state at the time of the last `setjmp` call. Procedure process is called to perform some processing and error checking. If an error is detected, a `longjmp` is performed, which changes the flow of execution directly to the first statement after the `setjmp`.

When should C be used?

The C language is good for embedded programming because it provides structure and flexibility without complex language restrictions.

What is the relationship between C and C++?

C++ is a hybrid object-oriented programming language that was originally implemented as macro-extension of C. Today, C++ stands by itself as a separately compiled language, although strictly speaking, C++ compilers should accept standard C code.

C++ exhibits all three characteristics of an object-oriented language. It promotes better software engineering practice through encapsulation and better abstraction mechanisms, such as inheritance, composition, and polymorphism, than does C.

When should C++ be used?

Significantly, more embedded systems are being constructed in C++ and many practitioners are asking this question. My answer to them is always "it depends." Choosing C in lieu of C++ in embedded applications is, roughly speaking, a tradeoff between a "lean and mean" C program that will be faster and easier to predict but harder to maintain and a C++

program that will be slower and unpredictable but potentially easier to maintain.

C++ still allows for low-level control (and not falling back to C features). For example, it can use inline methods rather than a runtime call. This kind of implementation is not completely abstract, nor completely low-level, but is acceptable in embedded environments.

What is the danger in converting my C code to C++?

There is some tendency to take existing C code and "objectify" it by wrapping the procedural code into objects with little regard for the best practices of object-orientation. This kind of approach should be avoided because it has the potential to incorporate all of the disadvantages of C++ and none of the benefits.

Can you tell me about Fortran?

Fortran* is the oldest high-order language (developed circa 1955) still in regular use today. Because in its earlier versions it lacked recursion and dynamic allocation facilities, more complex systems written in this language typically included a large portion of assembly language code to handle interrupts and scheduling. Communication with external devices was by memory-mapped I/O, Direct Memory Access (DMA), and I/O instructions. Later versions of the language included features such as re-entrant code, but even today a complex Fortran control system requires some assembly language code to accompany it.

To its detriment, Fortran is weakly typed, but because of the subroutine construct and the `if-then-else` construct, it can be used to design highly structured code. Fortran has no built-in exception handling or abstract data types.

Today, Fortran is still used to write some engineering applications because of its excellent handling of mathematical processing and because "old-time" engineers learned this language first. There is even a "new" language called F, which is a derivative of Fortran, less some esoteric and dangerous features. Many legacy systems, particularly in engineering applications, still can be found to have been written in "plain old" Fortran.

What about Java?

Java is an interpreted language; that is, the code compiles into machine-independent code which runs in a managed execution environment. This environment is a virtual machine (Figure 5.1), which executes "object" code instructions as a series of program directives.

* Although Fortran is an acronym for Formula Translator, it is often written as "Fortran" because the word has entered the mainstream in the same way that the acronyms "laser" and "sonar" have.

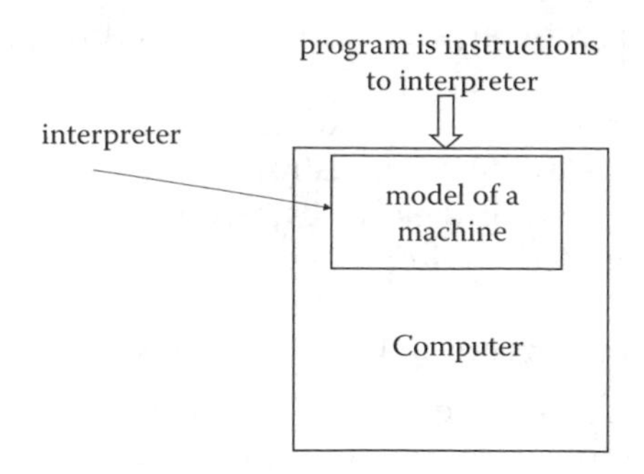

FIGURE 5.1
The Java interpreter as a model of a virtual machine.

The advantage of this arrangement is that the Java code can run on any device that implements the virtual machine. This "write once, run anywhere" philosophy has important applications in embedded and portable computing as well as in Web-based computing.

If Java is interpreted does that mean it cannot be compiled?

There are native-code Java compilers that allow Java to run directly "on the bare metal;" that is, the compilers convert Java directly to assembly code or object code.

What are some of the main features of Java?

Java is an object-oriented language that looks very much like C++. Like C, Java supports call-by-value, but the value is the reference to an object, which is in essence call-by-reference for all objects. Primitives are passed-by-value.

What are the differences between Java and C++?

The following are some features of Java that are different from C++ and that are of interest in engineering applications:

- There are no global functions or constants — everything belongs to a class.
- Arrays and strings have built-in bounds checking.
- All values are initialized and use special defaults if none are given.
- All classes ultimately inherit from the object class.
- Java does not support the goto statement. However, it supports labeled breaks, which can be used in the same way.
- Java does not support automatic type conversions (except where guaranteed safe).
- Types are all references to objects, except the primitive types (for example, integer, floating point, and Boolean).

Are there any well-known problems with using Java?

One of the best-known problems with Java involves its garbage collection utility. Garbage is memory that has been allocated but is unusable because of the loss of a pointer to it; for example, through the destruction of an object. The allocated memory must be reclaimed through garbage collection.

Garbage collection algorithms generally have unpredictable performance (although average performance may be known). The loss of determinism results from the unknown amount of garbage, the tagging time of the non-deterministic data structures, and the fact that many incremental garbage collectors require that every memory allocation or deallocation from the heap be willing to service a page-fault trap handler.

What about legacy code written in arcane languages such as BASIC, COBOL, and Scheme?

Code in many arcane languages still abounds. It is hard to explain how to deal with these situations except on a case-by-case basis.

What about Visual Basic?

There are different versions of Visual Basic, the oldest being derived from and resembling in many ways the ancient BASIC language. The newer versions exhibit strong object-oriented features. A great deal of production code has been written in Visual Basic, but as this is a proprietary language, it is beyond the scope of this text. Of course, there are many good texts available on this language.

What about scripting languages like Perl and Python?

Object-oriented scripting languages, such as Perl, Python, and Ruby, have become quite popular. Most scripting languages make it easier to write programs because of weak typing and their interpreted nature. The interpreted code leads to a shorter build cycle (high economy of compilation), which, in turn, allows for rapid prototyping and exploratory changes, a feature that is useful in agile development methodologies. However, generally speaking, it is more difficult to maintain production code that has been written in a scripting language [Chen et al. 2006].

5.2.4 Object-Oriented Languages — Fact and Fiction

It's pretty easy to learn an object-oriented language, isn't it?

It is a lot harder than you think. Mastering the syntax of the language is no harder than any other. However, experience has it that if you learned how to program in a procedural style language like Fortran, BASIC, or C, you will have a hard time understanding object-oriented principles. Conversely, software engineers who learn a pure object-oriented language like Java first usually have a difficult time using procedural languages. These engineers have an even harder time with primitive assembly language.

Can I learn how to program well in an object-oriented language like C++ or Java by taking a course?

Many self-styled pundits are out there teaching object-oriented programming who do not know the difference between a class and an object, or an interface and a parameter list. It takes many months and a great deal of practice to learn Java adequately. It can take years to become an expert.

I will be the first to admit that object technology is not my specialty. But I know just enough to know that it is relatively easy to write code in Java, C++, or C#,* but much harder to write good code that conforms to the best practices of object-oriented development.

A typical case in point is Joe, an experienced electrical engineer who took some Fortran and C programming classes 15 years ago when he was in college. He became pretty good at C programming because he was responsible for maintaining his company's library of digital signal processing code. About five years ago, his boss asked him to learn C++ and to convert his code to "object-oriented." Joe then essentially created a bunch of "God classes," huge classes that are not properly assigned responsibilities. Sure enough, the code compiled under the C++ compiler and ran. But the result was not object-oriented, just a poorly organized C program that happened to compile via a C++ compiler.

5.3 Software Construction Tools

What is the value proposition for using software construction tools?

Management of the software development phase can be greatly improved with version control or configuration management software, which regulates access to the various components of the system from the software library. Version control prevents multiple accesses to the same source code, provides mechanisms for tracking changes, and preserves version integrity. In the end, version control increases overall system reliability. There are number of freely available version control tools, such as CVS or SubVersion. CASE tools can also assist with software construction and in supporting many other software engineering activities.

What is a compiler?

A compiler translates a program from high-level source code language into relocatable object code or machine instructions or into an intermediate form such as an assembly language (a symbolic form of machine instructions). This process is broken into a series of phases, which will be described shortly.

* Pronounced "C sharp", C# is a close relative of C++ and runs only on Microsoft's .NET platform.

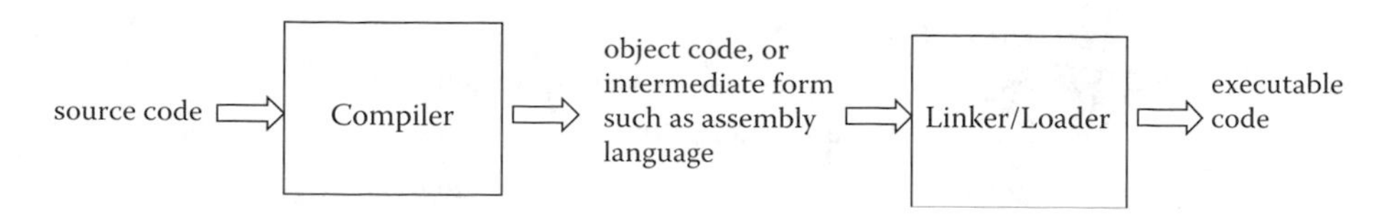

FIGURE 5.2
The compilation and linking process.

After compilation, the relocatable code output from the compiler is made absolute, and a program called a linker or linking loader resolves all external references. Once the program has been processed by the linker, it is ready for execution. The overall process is illustrated in Figure 5.2.

This process is usually hidden by any integrated development environment that acts as a front end to the compiler.

Can you describe further the compilation process?

Consider the widely used Unix-based C compiler. The Unix C compiler **cc** provides a utility that controls the compilation and linking processes. This compiler is also found in many Linux implementations. In particular, in Unix System V version 3 (SVR3), the **cc** program provides for the following phases of compilation:

- preprocessing
- compilation
- optimization
- assembly
- linking and loading

These phases are illustrated in Figure 5.3 and summarized in Table 5.1.

The preprocessing phase of the compiler, performed by the program **cpp**, takes care of things such as converting symbolic values into actual values and evaluating and expanding code macros. The compilation of the program, that is, the translation of the program from C language to assembly language, is performed by the program **ccom**.

Optimization of the code is performed by the program **c2**, and assembly or translation of the assembly language code into machine codes is taken

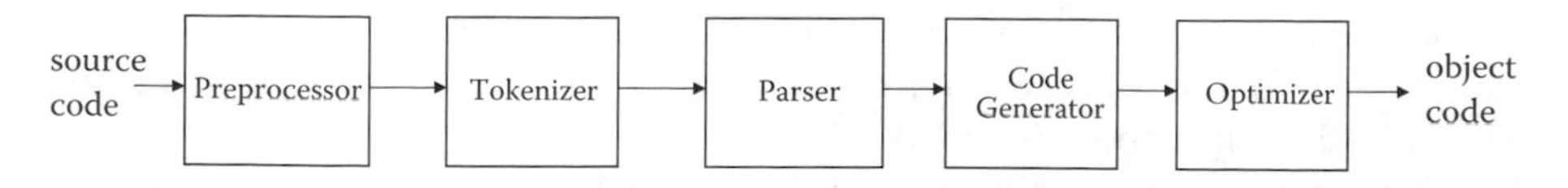

FIGURE 5.3
Phases of compilation provided by a typical compiler, such as C.

TABLE 5.1

Phases of Compilation and Their Associated Program for a C Compiler

Phase	Program
Preprocessing	cpp
Compilation	ccom
Optimization	c2
Assembly	as
Linking and loading	ld

care of by **as**. Finally, the object modules are linked together, all external references (for example, library routines) are resolved, and the program (or an image of it) is loaded into memory by the program **ld**. The executable program is now ready to run.

The fact that many of these phases can be bypassed or run alone is an important feature that can help in program debugging and optimization.

How do I deal with compiler warnings and errors?

One technique that can help is to redirect errors to a file. When a program with syntax errors is compiled, errors may be displayed to the screen too fast to read. These errors can be redirected to a file that can be looked at leisurely. This technique is called "redirecting standard error."

A difficulty that arises is that errors and warnings can be misleading; for example, an error reported in one place is indicative of a problem somewhere else. With experience, an error message can be associated with a particular problem, even when the error message appears far away from the error itself.

Another difficulty with errors and warnings is the cascade effect — the compiler finds one error and because no recovery is possible, many other errors follow. Again, experience with a particular compiler will enable the software engineer to more easily weed through the cascade of errors and find the root cause.

It is beyond the scope of this text to discuss the many kinds of program warnings and errors that the user can encounter in the course of compiling and linking programs. A complete discussion of this issue is best left to the reference book of the language in question.

Do you have any debugging tips?

Programs can be affected by syntactic or logic errors. Syntactic or syntax errors arise from the failure to satisfy the rules of the language. A good compiler will always detect syntax errors, although the way that it reports the error can often be misleading.

For example, in a C program a missing curly brace (}) may not be detected until many lines after it should have appeared, but in deeply nested arrays of curly braces it may be hard to see where the missing brace belongs.

In logic errors, the code adheres to the rules of the language but the algorithm that is specified is somehow wrong. Logic errors are more difficult to diagnose because the compiler cannot detect them. A few basic rules may help you find and eliminate logic errors.

- Document the program carefully. Ideally, each nontrivial line of code should include a comment. In the course of commenting, this may detect or prevent logical errors.
- Where a symbolic debugging is available, use steps, traces, breakpoints, skips, and so on to isolate the logic error (discussed later).
- In the case of a command line environment (such as Unix/Linux), output intermediate results at checkpoints in the code. This may help detect logic errors.
- In the case of an error, comment out portions of the code until the program compiles and runs. Add in the commented out code, one feature at a time, checking to see that the program still compiles and runs. When the program either does not compile or runs incorrectly, the last code you added is involved in the logic error.

Finding and eliminating logic errors is more art than science, and the software engineer develops these skills only with time and practice. In many cases, code audits or walkthroughs can be quite helpful in finding logic errors.

Is there any way to automatically debug code?

It is impossible to provide automatic logic validation and the compiler can check only for syntactical correctness. Many programming environments provide tools that are helpful in eliminating logical errors. For example, two tools (**lint** and **cb**) are associated with Unix and Linux. As its name implies, **lint** is a nit-picker that does checking beyond that of an ordinary compiler. For example, C compilers are often not particular about certain inconsistencies such as parameter mismatches, declared variables that are not used, and type checking. **lint** is, however. Often, very difficult bugs can be prevented or diagnosed by using **lint**.

The C beautifier, or **cb**, is used to transform a sloppy-looking program into a readable one. **cb** does not change the program code. Instead, it just adds tabs, line feeds, and spaces where needed to make things look nice. This is very helpful in finding badly matched or missing curly braces, erroneous if-then-else and case statements, and incorrectly terminated functions. As with **lint**, **cb** is run by typing **cb** and a file name at the command prompt.

Many open source integrated development environments (IDEs) have plug-ins that are the equivalent of or better than **lint** and **cb** and there are many other tools available, such as automatic refactoring engines, that can help improve code. Open source software is discussed in Chapter 8.

What are symbolic debuggers and how are they used?

Symbolic, or source-level, debuggers are software programs that provide the ability to step through code at either a macro-assembly or high-order language level. They are extremely useful in module-level testing. They are less useful in system-level debugging because the real-time aspect of the system is necessarily disabled or affected.

Debuggers can be obtained as part of compiler support packages or in conjunction with sophisticated logic analyzers. For example, **sdb** is a generic name for a symbolic debugger associated with Unix and Linux. **sdb** is a debugger that allows the engineer to single-step through source code language and view the results of each step. GNU has the GNU debugger, gdb, which can be used with GNU C and other languages such as GNU Java and Fortran.

In order to use the symbolic debugger, the source code must be compiled with a particular option set. This has the effect of including special run-time code that interacts with the debugger. Once the code has been compiled for debugging, then it can be executed "normally."

Can you give me an example of a debugging session?

In the Unix/Linux environment, the program can be started normally from the **sdb** debugger at any point by typing certain commands at the command prompt. However, it is more useful to single step through the source code. Lines of code are displayed and executed one at a time by using the "s" (for step) command. If the statement is an output statement, it will output to the screen accordingly. If the statement is an input statement, it will await user input. All other statements execute normally. At any point in the single-stepping process, individual variables can be set or examined.

There are many other features of **sdb**, such as breakpoint setting. In more integrated development environments, a graphical user interface (GUI) is also provided, but these tools essentially provide the same functionality.

What is a source code control system?

A source code control system places strict control over access to a software project's files to prevent conflicts such as multiple users editing a file simultaneously. In addition, source code control keeps an audit trail of changes and sets file access permissions. This kind of control is crucial when developing large programs for which many people will have access. Strict version control is quite important when handling programs consisting of a large number of files or when more than one individual is working on the same project.

For example, it would be disastrous if two programmers decided to modify the same source code file simultaneously — one set of changes would be lost. Similarly, suppose that a project consists of dozens of source files along with header and include files. If a header file is changed, every single source file using that header file must be recompiled or else very difficult-to-find bugs will be introduced.

Open source version control systems are available for free. These include SubVersion and CVS.

What is test driven design?

Test driven design (TDD) or test first coding is a code development approach in which the test cases for the code are written before the code is written. The advantage of this approach is that it forces the software engineer to think about the code in a very different way that involves focusing on "breaking down" the software. Many software engineers who use this technique report that while it is sometimes difficult to change their way of thinking, once the test cases have been written, it is actually easier to write the code. Furthermore, debugging becomes much easier because the unit level test cases have already been written.

Are there other tools that I can use, particularly when debugging embedded systems?

A number of hardware and software tools are available to assist in the validation of embedded systems. Test tools make the difference between success and failure, especially in deeply embedded systems.

For example, a multimeter can be used in the debugging of real-time systems to validate the analog input to or output from the system.

Oscilloscopes can be used for validating interrupt integrity, discrete signal issuance, and receipt, and for monitoring clocks. The more sophisticated storage oscilloscopes with multiple inputs can often be used in lieu of logic analyzers by using the inputs to track the data and address buses and synchronization with an appropriate clock.

Logic analyzers can be used to capture data or events, to measure individual instruction times, or to time sections of code. Programmable logic analyzers with integrated debugging environments further enhance the capabilities of the system integrator. Using the logic analyzer, the software engineer can capture specific memory locations and data for the purposes of timing or for verifying execution of a specific segment of code.

More sophisticated logic analyzers include built-in dissemblers and compilers for source-level debugging and performance analysis. These integrated environments typically are found on more expensive models, but they make the identification of performance bottlenecks particularly easy.

What are in-circuit emulators?

During module-level debugging and system integration of embedded systems, the ability to single-step the computer, set the program counter, and insert into and read from memory is extremely important. This capability, in conjunction with the symbolic debugger, is the key to the proper integration of embedded systems. In an embedded environment, however, an in-circuit emulator (ICE) provides this capability.

An ICE uses special hardware in conjunction with software to emulate the target CPU while providing the aforementioned features. Typically, the ICE plugs into the chip carrier or card slot normally occupied by the CPU. External wires connect to an emulation system. Access to the emulator is provided directly or via a secondary computer.

How are ICEs used?

ICEs are useful in software patching and for single-stepping through critical portions of code. ICEs are not typically useful in timing tests, however, because the emulator can introduce subtle timing changes.

What are software simulators and when are they used?

When integrating and debugging embedded systems, software simulators are often needed to stand in for hardware or inputs that do not exist or that are not readily available; for example, to generate simulated accelerometer or gyro readings where real ones are unavailable at the time.

The author of the simulator code has a task that is by no means easy. The software must be written to mimic exactly the hardware specification, especially in timing characteristics. The simulator must be rigorously tested; unfortunately, this is sometimes not the case. Many systems have been successfully validated and integrated with software simulators, only to fail when connected to the actual hardware.

When is hardware prototyping useful?

In the absence of the actual hardware system under control, simulation hardware may be preferable to software simulators. These devices might be required when the software is ready before the prototype hardware, or when it would be impossible to test the software on the actual hardware, such as in the control of a large nuclear plant.

Hardware simulators simulate real-life system inputs and can be useful for integration and testing but are not always reliable when testing the underlying algorithms; real data from live devices are needed.

What are integrated development environments?

Integrated development environments (IDEs) tie together various tools of the software production process through an easy to use GUI. IDEs can incorporate text editors, compilers, debuggers, coding standard enforcement, and much more. The most popular IDE includes the open source Eclipse.

What about other tools?

There are many other commercial and open source tools providing capabilities such as integrated document and software configuration control, testing management, stakeholder notification, distributed software build between cooperating PCs, and more.

5.4 Becoming a Better Code Developer

How can I become a better developer of code?

Coding software is not software engineering. The best code starts out with a good design. That being said, it is always important to improve your mastery of the programming language, just as you can improve your ability to write in English by better understanding the rules of the language, improving your vocabulary, and reading great works in the language. The best way to improve coding skills is to practice and read the appropriate literature. That raises one other point: one of the best ways to learn a programming language is to start by reading great code that someone else has written.

5.4.1 Code Smells

What is a code smell?

A code smell refers to an indicator of poor design or coding [Fowler 2000]. More specifically, the term relates to visible signs that suggest the need for refactoring. Code smells are found in every kind of system.

What is refactoring?

Refactoring refers to a behavior-preserving code transformation enacted to improve some feature of the software, which is evidenced by the code smell.

What are some code smells?

Table 5.2 summarizes a set of code smells originally described for object-oriented languages by Fowler [2000]. But many of these code smells are also found in procedural languages.

A few of these and several others not identified by Fowler that are found in procedural systems and described by Stewart [1999] will be discussed in terms of C, which has many constructs in common with C++, Java, and C#, so most readers should be able to follow the code fragments.

What is the conditional logic code smell?

These are excessive `switch`, `if-then`, and `case` statements and they are an indicator of bad design for several reasons. First, they breed code duplication. Moreover, the code generated for a case statement can be quite convoluted — for example, a jump through a register, offset by a table value. This mechanism can be time-consuming. Furthermore, nested conditional logic can be difficult to test, especially if it is nested due to the large number of logic paths through the code. Finally, the differences between best and worst case execution times can be significant, leading to highly pessimistic utilization figures.

TABLE 5.2

Some Code Smells and Their Indicators

Code Smell	Description
Alternative classes with different interfaces	Methods that do the same thing but have different signatures.
Conditional logic	`Switch` and `case` statements that breed duplication.
Data class	Classes with just fields, getters, setters, and nothing else.
Data clumps	Several data items found together in lots of places; should be an object.
Divergent change	When a class is changed in different ways for different reasons. Should be that each object is changed only for one type of change.
Duplicated code	The same code in more than one place.
Feature envy	Objects exit to package data with the processes used on that data.
Inappropriate intimacy	When classes spend too much time interacting with each others' private parts.
Incomplete library class	Poorly constructed library classes that are not reused.
Large class	A class that is doing too much.
Lazy class	A class that isn't doing enough to "pay" for itself.
Long method	Short methods provide indirection, explanation, sharing, and choosing.
Long parameter list	Remnant of practice of using parameter lists vs. global variables.
Message chains	The client is coupled to the structure of the navigation: getA().getB().getC().getD().getE().doIt().
Middle man	When a class delegates too many methods to another class.
Parallel inheritance hierarchies	Special case of shotgun surgery. Need to ensure that instances of one hierarchy refer to instances of the other.
Primitive obsession	Aversion to using small objects for small tasks.
Refused bequest	When a child only wants part of what is coming from the parent.
Shotgun surgery	Every time you make a change you have to make a lot of little changes. Opposite of divergent change.
Speculative generality	Hooks and special cases to handle things that are not required (might be needed someday).
Tell-tale comments	Comments that exist to hide/explicate bad code.
Temporary field	Instance variables that are only set sometimes; you expect an object to use all of its variables.

Source: Fowler, M., *Refactoring*, Addison-Wesley, Boston, MA, 2000.

Conditional logic needs to be refactored, but there is no silver bullet here. In object-oriented languages, the mechanisms of polymorphism or composition can be used. In procedural languages, sometimes mathematical identities can be used to improve the efficiency of the code.

What are data clumps?

Several data items found together in lots of places are known as data clumps. For example, in the procedural sense, data clumps can arise in C from too much configuration information in `#include files`. Stewart [1999] notes that this situation is unhealthy because it leads to increased development and maintenance time and introduces circular dependencies that make reuse difficult. He suggests that to refactor, each module be defined by two files, `.h` and `.c`, with the former containing only the information that is to be exported by the module and the latter containing everything that is not exported.

A second manifestation of the data clump smell has to do with excessive use of the `#define` statement that propagates throughout the code. Suppose these `#defines` are expanded in 20 places in the code. If, during debugging, it is desired to place a patch over one of the `#defines`, it must be done in 20 places. To refactor, place the quantity in a global variable that can be changed in one place only during debugging.

Why are delays as loops bad?

Stewart [1999] suggests another code smell involving timing delays implemented as **`while`** loops with zero or more instructions. These delays rely on the overhead cost of the loop construct plus the execution time of the body to achieve a delay. The problem is that if the underlying architecture or characteristics of instruction execution (e.g., memory access time change) changes, then the delay time is inadvertently altered.

To refactor, use a mechanism based on a timer facility provided by the operating system that is not based on individual instruction execution times.

What are dubious constraints?

This code smell is particularly insidious in embedded systems where response time constraints have a questionable or no attributable source. In some cases, systems have deadlines that are imposed on them that are based on nothing more than guessing or on some forgotten and since eliminated requirement. The problem in these cases is that the undue constraints may be placed on the systems.

For example, suppose the response time for an event is 30 ms but no one knows why. Similarly, more than one reason given for the constraints in comments or documentation indicates a traceability conflict, which hints at other problems. This is a primary lesson in embedded systems design to understand the basis and nature of the timing constraints, so that they can be relaxed if necessary.

It is typical, in studying embedded systems, to consider the nature of time because deadlines are instants in time. But the question arises, "where do the deadlines come from?" Generally speaking, deadlines are based on the underlying physical phenomena of the system under control. For example, in animated displays, real-time images must be updated at approximately 30 frames per second to provide a continuous image because the human eye can resolve updating at a slower rate. In navigation systems, accelerations must be read at a rate that is based on the top speed of the vehicle, and so on.

In any case, to remove the code smell, some detective work is needed to discover the true reason for the constraint. If it cannot be determined, then the constraint could be relaxed and the system redesigned accordingly.

What is the duplicated code smell?

Obviously, duplicated code refers to the same or similar code found in more than one place. It has an unhealthy impact on maintainability (the same change has to be propagated to each copy), and it also adversely affects memory utilization.

To refactor, assign the code to a single common code unit via better application of information hiding.

While it is too easy to mock the designers of systems that contain duplicated code, it is possible that the situation arose out of a real need at the time. For example, duplicated code may have been due to legacy concerns for performance where the cost of the procedure call added too much overhead in a critical instance. Alternatively, in older version of languages such as Fortran that were not reentrant, duplicated code was a common means for providing utilities to each cycle in the embedded system.

What are generalizations based on a single architecture?

Stewart [1999] suggests that writing software for a specific architecture, but with the intent to support easy porting to other architectures later, can lead to over-generalizing items that are similar across architectures, while not generalizing some items that are different. He suggests developing the code simultaneously on multiple architectures and then generalizing only those parts that are different. He also suggests choosing three or four architectures that are very different in order to obtain the best generalization. Presumably, such an approach suggests the appropriate refactoring.

What are the large method, large class, and large procedure code smells?

Fowler [2000] describes two code smells, long method and large class, which are self-evident. In the procedural sense, the analogy is a large procedure. Large procedures are anathema to the divide-and-conquer principle of software engineering and need to be refactored by re-partitioning the code appropriately.

What are lazy methods and lazy procedure?

A lazy method is one that does not do enough to justify its existence. The procedural analogy to the lazy class/method code smells is the lazy procedure. In a real-time sense, a procedure that does too little to pay for the overhead of calling the procedure needs to be eliminated by removing its code to the calling procedure or redefining the procedure to do more.

What is the long parameter list code smell and how can it be refactored?

Long parameter lists are an unwanted remnant of the practice of using parameter lists to avoid the use of global variables. While clearly well-defined interfaces are desirable, long parameter lists can cause problems in embedded systems if interrupts are disabled during parameter passing. In this case, overly long interrupt latencies and missed deadlines are possible.

The long parameter list code smell can be refactored by passing a pointer to one or more data structures that contain aggregated parameters, or by using global variables.

What are message chains?

Fowler [2000] describes message chains as a bad smell in OOD. For example, a message chain occurs if the client is coupled to the structure of the navigation: `getA().getB().getC().getD().getE().doIt();`.

The procedural analogy to message chains might be an overly long sequence of procedure calls that could be short circuited or replaced by a more "reasonable" sequence of calls. The problem in the case of long chains is that the overhead of calling procedures becomes significant, interrupts may be disabled during parts of the procedure calls (e.g., for parameter passing), and the long sequence of calls may indicate inefficient design.

What is message passing overload?

Stewart [1999] describes the excessive use of message passing for synchronization as another unwanted practice. He notes that this practice can lead to unpredictability (because of the necessary synchronization), the potential for deadlock, and the overhead involved.

He suggests that the refactoring is to use state-based communication via shared memory with structured communication.

What is the one big loop code smell and how is it refactored?

Cyclic executives are non-interrupt driven systems that can provide the illusion of simultaneity by taking advantage of relatively short processes on a fast processor in a continuous loop. For example, consider the set of self-contained processes `Process1` through `Processn` in a continuous loop as depicted in the following:

```
for(;;)      {                    /* do forever  */
              Process1();
              Process2();
              ...
              Processn()
             }
      }
```

Stewart [1999], who calls the cyclic executive "one big loop," notes that there is little flexibility in the cyclic executive and that only one cycle rate is created. However, different rate structures can be achieved by repeating a task in the list. For example, in the following code

```
for(;;)      {                          /* do forever  */
               Process1();
               Process2();
               Process3();
               Process3();
             }
        }
```

`Process3` runs twice as frequently as `Process1` or `Process2`.

Moreover, the task list can be made flexible by keeping a list of pointers to processes, which can be managed by the "operating system" as tasks are created and completed. Intertask synchronization and communication can be achieved through global variables or a parameter list.

Generally, however, the "big loop" structure really is a bad smell unless each process is relatively short and uniform in size. The refactoring involves rebuilding the executive using an interrupt scheme such as rate-monotonic or round-robin.

What is shotgun surgery?

Shotgun surgery is a very common code smell related to the phenomenon that every time you make a change you have to make many little changes. This is another example of poor application of information hiding, which suggests the refactoring.

What is speculative generality?

Speculative generality relates to hooks and special cases that are built into the code to handle things that are not required (but might be needed someday).

Embedded systems are no place to build in hooks for "what-if" code. Hooks lead to testing anomalies and possible unreachable code. Therefore, the refactoring is to remove hooks and special cases that are not immediately needed.

What are tell-tale comments?

The tell-tale comment problem appears in all kinds of systems. Comments that are excessive, or which tend to explicate the code beyond a reasonable level, are often indicators of some serious problem. Comments such as "do not remove this code," or "if you remove this statement the code doesn't work, I don't know why" are not uncommon. Humor in comment statements can sometimes be a glib way to mask the fact that the writer doesn't know what he is doing. Oftentimes these kinds of statements indicate that there are underlying timing errors.

In either case, the refactoring involves rewriting the code so that an overly long explicating comment is not necessary.

What are temporary fields and how are they refactored?

Temporary fields are instance variables that are only set sometimes; you expect an object to use all of its variables. In the procedural sense, this could be seen in the case of a `struct` with fields that are unused in certain instances. The refactoring is to use an alternative data structure. For example, in C you could use a `union` to replace the `struct` .

What is the unnecessary use of interrupts code smell?

Stewart [1999] also suggests that indiscriminate use of interrupts is a bad code smell. Interrupts can lead to deadlock, priority inversion, and inability to make performance guarantees [Laplante 2004b].

Interrupt-based systems should be avoided, although it has been noted that avoidance is possible only in the simplest of systems where real-time multitasking can be achieved with coroutines and cyclic executives implemented without interrupts. When interrupts are required to meet performance constraints, rate-monotonic or earliest deadline first scheduling should be used.

How can I improve the run-time performance of the code I write?

Many of these improvements can be had by refactoring the code smells just discussed. However, the main thing you can do to improve code execution performance, particularly in embedded systems, is to understand the mapping between high-order language input and assembly language output for a particular compiler. This understanding is essential in generating code that is optimal in either execution time or memory requirements. The easiest and most reliable way to learn about any compiler is to run a series of tests on specific language constructs.

For example, in many C compilers the `case` statement is efficient only if more than three cases are to be compared; otherwise, nested `if` statements should be used. Sometimes the code generated for a `case` statement can be quite convoluted; for example, a jump through a register offset by the table value. This sequence can be time-consuming.

It has already been mentioned that procedure calls are costly in terms of passing of parameters via the stack. The software engineer should determine whether the compiler passes the parameters by byte or by word.

Other language constructs that may need to be considered include:

- Use of `while` loops vs. `for` loops or `do-while` loops.
- When to "unroll" loops; that is, to replace the looping construct with repetitive code (thus saving the loop overhead as well as providing the compiler the opportunity to use faster machine instructions).

- Comparison of variable types and their uses (e.g., when to use short integer in C vs. Boolean, when to use single precision vs. double precision floating point, and so forth).
- Use of in-line expansion of code via macros vs. procedure calls.

This is by no means an exhaustive list of tips.

5.4.2 Coding Standards

What are coding standards?

Coding standards are different from language standards. A language standard (e.g., ANSI C*) embodies the syntactic rules of the language. A program violating those rules will be rejected by the compiler. Conversely, a coding standard is a set of stylistic conventions. Violating the conventions will not lead to compiler rejection. In another sense, compliance with language standards is mandatory while compliance with coding standards is voluntary.

How can coding standards help improve my code?

Adhering to language standards fosters portability across different compilers and, hence, hardware environments. Complying with coding standards will not likely increase portability, but rather in many cases will increase readability and maintainability. Some even contend that the use of coding standards can enhance reliability. Coding standards may also be used to promote improved performance by encouraging or mandating the use of language constructs that are known to generate code that is more efficient. Many agile methodologies, for example, eXtreme Programming, embrace coding standards.

What do coding standards look like?

Coding standards involve standardizing some or all of the following elements of programming language use:

- Standard or boilerplate header format.
- Frequency, length, and style of comments.
- Naming of classes, methods, procedures, variable names, data, file names, and so forth.
- Formatting of program source code including use of white space and indentation.
- Size limitations on code units including maximum and minimum lines of code, number of methods, and so forth.
- Rules about the choice of language construct to be used; for example, when to use `case` statements instead of nested `if-then-else` statements.

This is just a partial list.

---

* The American National Standards Institute Standard for the C language.

What is the benefit of coding standards?

Close adherence to a coding standard can make programs easier to read and understand and likely more reusable and maintainable.

Which coding standard should I use?

Many different standards for coding are language-independent or language-specific. Coding standards can be team-wide, company-wide, or user-group specific. For example, the Gnu software group has standards for C and C++, or customers can also require conformance to a specific standard that they own. Still other standards have become public domain.

One example is the Hungarian notation standard, named in honor of Charles Simonyi, who is credited with first promulgating its use. Hungarian notation is a public domain standard intended to be used with object-oriented languages, particularly C++. The standard uses a complex naming scheme to embed type information about the objects, methods, attributes, and variables in the name. Because the standard essentially provides a set of rules about naming variables, it can be used with other languages such as C++, Ada, Java, and even C.

Are there any drawbacks to using coding standards?

First, let me say that I believe that you should always follow a coding standard, despite any difficulties they might introduce or shortcomings they might have. One problem with standards, however, is that they can promote very mangled variable names. In other words, the desire to conform to the standard is greater than creating a particularly meaningful variable name.

Another problem is that the very strength of coding standards can be their undoing. For example, in Hungarian notation what if the type information embedded in the object name is wrong? There is no way for a compiler to check this. There are commercial rule wizards that can be tuned to enforce the coding standards, but they must be tuned to work in conjunction with the compiler.

When should the coding standard be adopted?

Adoption of coding standards is not recommended mid-project. It is much easier to conform at the start of the project than to be required to change existing code.

5.5 Further Reading

Cardelli, L. Bad engineering properties of object-oriented languages, *ACM Comp. Surveys*, 28A(4), 150–158, 1996.

Chen, Y., Dios, R., Mili, A., Wu, L., and Wang, K., An empirical study of programming language trends, *IEEE Software*, 22(3), 72–78, 2006.

Fowler, M., *Refactoring,* Addison-Wesley, Boston, MA, 2000.

Gamma, E., Helm, R., Johnson, R., and Vlissides, J., *Design Patterns: Elements of Reusable Object-Oriented Software,* Addison-Wesley, Boston, MA, 1995.

Kernighan, B.W. and Ritchie, D.M., *The C Programming Language,* 2nd ed., Prentice Hall, Englewood Cliffs, NJ, 1988.

Laplante, P.A., *Software Engineering for Image Processing Systems,* CRC Press, Boca Raton, FL, 2004a.

Laplante, P.A., *Real-Time Systems Design and Analysis: An Engineer's Handbook,* 3rd ed., IEEE Press/John Wiley & Sons, New York , 2004.

Louden, K.C., *Programming Languages: Principles and Practice,* 2nd ed., Thomson Course Technology, Boston, MA, 2002.

McConnell, S., *Code Complete,* 2nd ed., Microsoft Press, Redmond, WA, 2004.

Meyer, B., *Object-Oriented Software Construction,* 2nd ed., Prentice Hall, Englewood Cliffs, NJ, 2000.

Sebesta, R.W., *Concepts of Programming Languages,* 7th ed., Addison-Wesley, Boston, MA, 2006.

Stewart, D.B. Twenty-five most common mistakes with real-time software development, Class #270, *Proc. 1999 Embedded Syst. Conf.*, San Jose, CA, 1999.

6

Software Quality Assurance

Outline

- Quality models and standards
- Software Testing
- Metrics
- Fault tolerance
- Maintenance and reusability

6.1 Introduction

"In the beginning of a malady it is easy to cure but difficult to detect, but in the course of time, not having been either detected or treated in the beginning, it becomes easy to detect but difficult to cure [Machiavelli, 1513]." Machiavelli's sentiments are precisely those that must abound in a software enterprise seeking to produce a quality product. Apparently, though, this sentiment is not prevalent. A 2002 study by the National Institute of Standards Technology (NIST) estimated that software errors cost the U.S. economy $59.5 billion each year. The report noted that software testing could have reduced those costs to about $22.5 billion. Of the $59.5 billion, users paid for 64% and developers paid for 36% of the cost [NIST 2002].

In order to achieve the highest levels of software quality, there are several things you have to do. First, you need to have in place a system that will foster the development of quality software. This is what the CMM-I (to be discussed later) is all about. Incorporated in that quality software system is rigorous, life cycle testing.

In order to attain any of the desirable software qualities, you have to have a way to measure them through metrics. Finally, you can improve the reliability of the software through fault-tolerant design. This chapter discusses all of these aspects of software quality.

6.2 Quality Models and Standards

What is software quality?

There are many ways that stakeholders might perceive software quality, all based on the presence or absence to a certain degree of one attribute or another. Some formal definitions are appropriate, however, as defined in the ***ISO*** *8402:2000 Quality Management and Quality Assurance — Vocabulary standard*.

Quality is the totality of features and characteristics of a product or service that bear on its ability to satisfy stated or implied needs. A quality policy describes the overall intentions and direction of an organization with respect to quality, as formally expressed by top management. *Quality management* is that aspect of overall management function that determines and implements quality policy. Finally, a quality system is the organizational structure, responsibilities, procedures, processes, and resources for implementing quality management.

As it turns out, one can undermine any of these definitions with rhetorical arguments, but for our purposes, they are useful working definitions. These are broad definitions of quality for every kind of product, not just software. There are other ways to look at software quality.

For example, Voas and Agresti [2004] propose that quality is comprised of a set of key behavioral attributes such as:

- reliability (R)
- performance (P)
- fault tolerance (F)
- safety (Sa)
- security (Se)
- availability (A)
- testability (T)
- maintainability (M)

along with a dash of uncertainty. They further suggest that quality has a slightly different meaning for each organization, and perhaps each application, which can be represented by a weighted linear combination of these behavioral attributes, namely,

$$Q = w_1R + w_2P + w_3F + w_4Sa + w_5Se + w_6A + w_7T + w_8M + \text{uncertainty}$$

Munson [2003] discusses quality as a set of objectives:

- Learn to measure accurately people, processes, products, and environments.
- Learn to do the necessary science to reveal how these domains interact.
- Institutionalize the process of learning from past mistakes.
- Institutionalize the process of learning from past successes.
- Institutionalize the measurement and improvement process.

All of these different takes on quality are compatible and largely consistent with the prevailing quality models.

What is a quality model?

A quality model is a system for describing the principles and practices underlying software process maturity. A quality model is different from a life-cycle model in that the latter is used to describe the evolution of code from conception through delivery and maintenance. A quality model is intended to help software organizations improve the maturity of their software processes as an evolutionary path from *ad hoc,* chaotic processes to mature, disciplined software processes. The most famous and widely employed quality model in the software industry is the capability maturity model (CMM).

What is the capability maturity model?

The CMM is a software quality model consisting of five levels. Predictability, effectiveness, and control of an organization's software processes are believed to improve as the organization moves up these five levels. While not truly rigorous, there is some empirical evidence that supports this position.

The CMM for software is not a life-cycle model, but rather a system for describing the principles and practices underlying software process maturity. CMM is intended to help software organizations improve the maturity of their software processes in terms of an evolutionary path from ad hoc, chaotic processes to mature, disciplined software processes.

What is the history of the CMM?

CMM had its foundations in work begun in 1986 at the U.S. DoD to help improve the quality of the deliverables produced by government software contractors. The work was commissioned through the MITRE Corporation, but later moved to the Software Engineering Institute (SEI) at Carnegie Mellon University. Watts Humphrey was the initial author, and then Mark Paulk, Bill Curtis, and others took the lead in CMM's development [Paulk et al. 2005]. CMM borrows heavily from general Total Quality Management (TQM) and the work of Philip Crosby.

What is the capability maturity model integration?

The capability maturity model integration (CMM-I) is a more generic version of the CMM that is suitable for other endeavors beside software. CMM-I consists of three parts: one for software (SW-CMM), one for systems engineering that includes integrated product and process development (SE-CMM), and one that includes some acquisition aspects (IPD-CMM). The CMM-I product suite comprises multiple integrated models, courses, and an assessment method.

For convenience, I will refer to CMM and CMM-I interchangeably.

What is the CMM "maturity suite?"

This consists of a set of maturity models for different aspects of the software enterprise. It includes

- Software CMM (SW)
- Personal Software Process (PSP)
- Team Software Process (TSP)
- People CMM (P)
- Software Acquisition CMM (SA)
- System Engineering CMM (SE)
- Integrated Product Development CMM (IPD)
- CMM Integration (CMM-I)

The CMM-I is comprised of the SW, SE, and IPD CMM suites taken together.

What are the levels of CMM?

The levels are:

Initial
Repeatable
Defined
Managed
Optimizing

I will discuss each of these in turn.

What are the components of each CMM level?

The CMM consists of the five maturity levels just mentioned. Each maturity level is a well-defined evolutionary stage of a mature software process (except for Level 1 where the processes are *ad hoc*). The maturity levels indicate process capabilities, which describe the range of expected results that can be achieved by following a software process. Maturity levels also contain key process areas, which identify related activities that help achieve a set of important goals for that level. Key process areas are organized by common features. The common features are an attribute that indicate whether the implementation and institutionalization of a key process area is effective, repeatable, and lasting. Finally, the common features contain key practices that describe the infrastructure and activities that contribute most to the effective implementation and institutionalization of the key process area.

Should there be a Level 0 in the CMM?

Level 0 is not officially recognized, but it is sometimes characterized as "chaos."

What is CMM Level 1?

Level 1 is called the "initial" level. In Level 1, the organization is characterized by software processes that are *ad hoc* and chaotic. Level 1 is not well defined and organizations may have some aspects at Level 1 while others are further evolved. Most organizations probably exhibit some "pockets" of Level 1 behavior.

What are some of the characteristics of Level 1 organizations?

Level 1 organizations are not as chaotic as you might think. There may be evolved business processes, which are possibly holdovers from TQM or Malcolm Baldridge quality initiatives, but they are working against each other. These organizations can still have strong successes but success might not be repeatable because each team tackles projects in different ways each time.

In Level 1 organizations, there is little or no measurement and while some time/cost estimates are accurate, many are far off. In fact, success in these settings comes not from a well-managed software operation but from smart people doing the right things. When this is the case, there are frequent crises and firefighting and it is very hard for the organization to recover when good people leave.

What is CMM Level 2?

Level 2 is the "repeatable" level. Here, basic project management processes are established to track cost, schedule, and functionality. The necessary process discipline is in place to repeat earlier successes on projects with similar applications. Key process focus on:

- requirements management
- software project planning
- software project tracking
- oversight
- software subcontract management
- software quality assurance
- software configuration management

Conventional wisdom is that it takes a committed organization approximately 18 months to advance from Level 1 to Level 2.

What is CMM Level 3?

CMM Level 3 is the "defined" level. Here the organization has achieved Level 2 maturity plus the software process for both management and engineering activities is documented, standardized, and integrated into a standard software process for the organization. All projects use an approved, customized version of the organization's standard software process for developing and maintaining software.

The key process areas at this level address both project and organizational issues, as the organization establishes an infrastructure that institutionalizes effective software engineering and management processes across all projects. Key process areas are:

- organization process focus
- organization process definition
- training program
- integrated software management
- software product engineering
- inter-group coordination
- peer reviews

Characteristically, all projects use an approved, customized version of the organization's standard software process for developing and maintaining software.

What is CMM Level 4?

Level 4 is the "managed" level and it consists of all of the Level 3 characteristics. Here, detailed measures of the software process and product quality are collected. Both the software process and products are quantitatively understood and controlled.

The key process areas here focus on establishing a quantitative understanding of both the software process and the software work products being built. They are quantitative process management and software quality management.

What is CMM Level 5?

Level 5 is the "optimizing" level. It consists of Level 4 characteristics plus evidence of continuous process improvement enabled by quantitative feedback from the process and from piloting innovative ideas and technologies. The key process areas at this level cover the issues that both the organization and the projects must address to implement continual, measurable software process improvement. They are defect prevention, technology change management, and process change management.

How many organizations are currently CMM-I certified?

Level 1 organizations are surely the most common. Only a small percentage of software organizations attempt a CMM assessment, so there is no way of knowing how many are at a given level. But a hint is given by some proportionate statistics from a 2003 survey of 70 organizations shown in Figure 6.1 [Zubrow 2003].

Zubrow's estimates, based on his sample, show that about 40% of the organizations achieve Level 3 upon first appraisal.

How can my organization use the CMM?

There are several ways that the CMM-I certification can be used to improve your organization's overall software practice. For example, it can hire an officially certified CMM assessor to conduct an initial evaluation, then work to help the organization achieve a certain level of maturity (Level 3 is often the target).

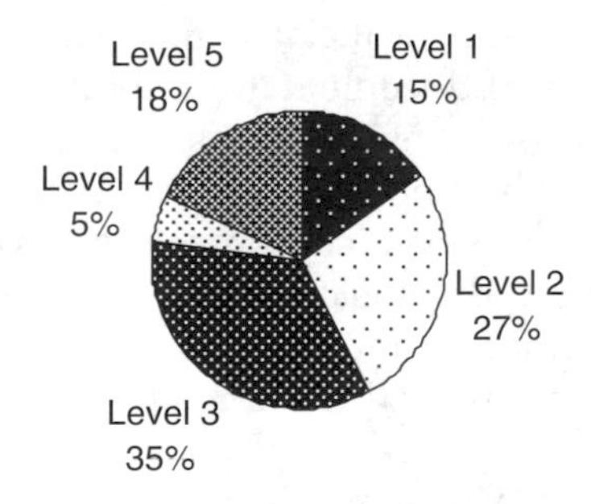

FIGURE 6.1
Maturity profile of 70 CMM certified organizations surveyed. (From Zubrow, D., Current Trends in the Adoption of the CMMI® Product Suite, *compsac, 27th Annu. Intl. Comp. Software Appl. Conf.*, 2003, pp. 126–129.)

Reaching a certain level is usually a prerequisite to bid on certain government software contracts or to work with other customers and partners that value such certification.

Likewise, your company may judge that a certain level of maturity is a prerequisite for its partners. Some development house, particularly those oversees, use CMM certification as a "Good Housekeeping Seal of Approval" to impress potential clients.

If your company prefers, it can send its own staff to official CMM training (for example, through the SEI) to evangelize CMM throughout your organization. This is especially cost effective for a large organization where software process improvements will have a big payoff.

Finally, and I think most commonly, your organization can use CMM as a set of suggestions and apply them as appropriate.

Why customize CMM this way?

For some, especially smaller organizations, the cost of certification and maintaining the practices can be expensive. To others, the CMM might just seem like a word game. However, if you get beyond the verbiage, CMM provides important guidance on how to build software.

Are there "conventional" objections to using the CMM?

There a variety of arguments proffered against seeking CMM certification for a software organization. The most obvious objection is cost — it will take significant time and money to prepare for certification and to move up the various levels. The counter-argument to this objection, of course, is that the cost is worth it. In my opinion, it is very hard to use cost justification for either side of this argument because of the many variables that govern success at one organization and failure at another. It is just impossible to impute the experiences in one case to another.

Some argue that CMM is an ends to a means but not a means (for developing quality software) itself. Indeed, some cynics believe that a CMM environment gives a somewhat artificial stamp of approval. They justify this argument because the CMM documentation, though lengthy, does not provide any particular recipes for any of the key practices. It just says which key practices should exist, not how to perform them.

Consider Jane, who works at a company that recently achieved CMM Level 3. She was tired and disillusioned. Why? She wanted her company to say, "Let's work together to improve our software processes." Instead, they said, "Now that we achieved Level 3, let's get back to what we were doing."

Another objection to CMM is that it is not helpful in a crisis. Indeed, that is true, but that is not the purpose of achieving any CMM level. West [2004] uses the analogy of someone pledging to diet and exercise as he is having a heat attack. Of course, by then it is too late; the person should have dieted and exercised long before as a preventative measure. So, too, it is with CMM.

Many of the conventional objections are borne from myths that have been propagated by naysayers.

What are some of the CMM-I and process implementation myths?

West [2004] lists quite a few of these myths:

- Myth: CMM or CMM-I gives organizations requirements for developing successful programs.
- Fact: It helps describe the conditions under which successful software development programs are likely to exist.
- Myth: Having higher maturity levels ensures a software or systems organization will be successful.
- Fact: Not necessarily. You can get bad food at a clean restaurant. However, you are not likely to get sick from that food.
- Myth: Before implementing CMM or CMM-I, organizations are usually in total chaos.
- Fact: Not so. There are plenty of good software development organizations that have chosen not to seek CMM certification. Moreover, even those at Level 1 can have success.
- Myth: The primary and best reason for process improvement is to achieve maturity levels.
- Fact: We've debunked this notion already.
- Myth: CMM will fix all your software development problems.
- Fact: Obviously untrue; there is no "silver bullet."
- Myth: Model-based process improvement does not affect what I do.
- Fact: Again, this is obviously not correct.
- Myth: Implementing process improvement based on CMM is rocket science and only a few geniuses understand it.
- Fact: The CMM concepts are relatively straightforward. Eat right, get plenty of sleep, brush your teeth three times a day. But these practices can be hard to live by in reality.

6.2.1 Other Quality Standards and Models

What is ISO 9000?

ISO 9000 is a generic, worldwide standard for quality improvement. The International Standards Organization owns the standard.

Collectively described in five standards, ISO 9000 through ISO 9004, ISO 9000 was designed to be applied in a wide variety of manufacturing environments.

ISO 9001 through ISO 9004 apply to enterprise according to the scope of their activities. ISO 9004 and ISO 9000-X family are documents that provide guidelines for specific applications domains.

ISO 9000-3 (1997) is essentially an expanded version of ISO 9001 with added narrative to encompass software. The standard is widely adopted in Europe and increasingly in the U.S. and Asia.

What are ISO 9000-3 principal areas of quality focus?

They are:

- management responsibility
- quality system requirements
- contract review requirements
- product design requirements
- document and data control
- purchasing requirements
- customer supplied products
- product identification and traceability
- process control requirements
- inspection and testing
- control of inspection, measuring, and test equipment
- inspection and test status
- control of nonconforming products
- corrective and preventive actions
- handling, storage, and delivery
- control of quality records
- internal quality audit requirements
- training requirements
- servicing requirements
- statistical techniques

What does ISO 9000-3 look like?

It is very short (approximately 30 pages) and very high level (see Figure 6.2 for an excerpt).

Are there any similarities between ISO 9000-3 and CMM?

CMM-I is a model for operational excellence, while ISO 9000-3 is a standard for quality software systems. While both are abstractions, the level of detail is much different; ISO 9000-3 is approximately 30 pages long while CMM-I is approximately 700 pages long.

ISO 9000-3	**4.4 Software development and design**
4.4.1 General	Develop and document procedures to control the product design and development process. These procedures must ensure that all requirements are being met.
Software development	Control your software development project and make sure that it is executed in a disciplined manner.

- Use one or more life cycle models to help organize your software development project.
- Develop and document your software development procedures. These procedures should ensure that:
  - Software products meet all requirements.
  - Software development follows your:
    - Quality plan.
    - Development plan.

FIGURE 6.2
Excerpt from ISO 9000-3: 4.4 Software development and design.

Are ISO 9000-3 and CMM compatible?

They are clearly not the same, but they are compatible. Both identify best practices but neither is explicit about how to do things. ISO 9000-3 defines minimum standards for organizations, while CMM-I defines different levels of operational excellence.

How does ISO/IEC 12207 help promote software quality?

This standard, which was briefly introduced in Chapter 2, is contained in a relatively high-level document used to help organizations refine their processes, by defining compliance as the performance of those processes, activities, and tasks. Hence, it allows an organization to define and meet its own quality standards. Because it is so general, organizations seeking to apply 12207 need to use additional standards or procedures that specify those details.

What is Six Sigma?

Developed by Motorola, Six Sigma is a management philosophy based on removing process variation. Six Sigma focuses on the control of a process to ensure that outputs are within six standard deviations (six sigma) from the mean of the specified goals. Six Sigma is implemented using define, measure, improve, analyze, and control (DMIAC).

Define means to describe the process to be improved, usually through some sort of business process model. *Measure* means to identify and capture relevant metrics for each aspect of the process model. The goal-question-metric paradigm is helpful in this regard. *Improve* obviously means to change some aspect of the process so that beneficial changes are seen in the associated metrics, usually by attacking the aspect that will have the highest payback. *Analyze* and *control* mean to use ongoing monitoring of the metrics to continuously revisit the model, observe the metrics, and refine the process as needed.

Some organizations use Six Sigma as part of their software quality practices. The issue here, however, is in finding an appropriate business process model for the software production process that does not devolve into a simple, and highly artificial, waterfall process.

What is the relationship between Six Sigma and the CMM?

The two can be used in a complementary fashion; Six Sigma is weak on improvement infrastructure while CMM-I is strong in this regard.

Six Sigma is more process-yield-based than CMM-I, so CMM-I process areas can be used to support DMIAC (e.g., by encouraging measurement). While CMM-I identifies activities, Six Sigma helps optimize those activities. Six Sigma can also provide specific tools for implementing CMM-I practices (e.g., estimation and risk management).

What is the IT infrastructure library?

The IT infrastructure library (ITIL) is a worldwide standard for IT service management. Originated in the U.K. (and "owned" by its Office of Government Commerce), ITIL has standards for

- service support
- service delivery
- infrastructure management
- application management
- planning to implement
- business perspective

"Dashboard" tools are available for each of these standards.

Can ITIL help with software quality programs?

Some companies find ITIL to be a useful framework for software quality management because it is non-proprietary, platform-independent, and adaptable. At this writing, ITIL is used as the basis for the Microsoft Operations Framework (MOF) and the Hewlett Packard (HP) IT service management reference model.

How does ITIL help with software quality management?

ITIL helps create comprehensive, consistent, and coherent codes of best practice for quality IT service management, promoting business effectiveness in the use of IT. ITIL also encourages the private sector to develop ITIL-related services and products, training, consultancy, and tools.

Can anything bad come of a software quality initiative?

Yes. According to West [2004], "slash-and-burn improvement" can occur. This approach means throwing away the good practices (and perhaps people) along with the bad; for example, when it is determined (or assumed) that an organization is at CMM Level 1, and the decision is made to tear down the existing software culture and start from scratch. The negative effects of slash-and-burn process improvement are not always easy to detect [West 2004].

What are the symptoms of slash-and-burn approaches?

There are many possible signs. West [2004] gives the following:

- A process improvement specialist (consultant) cannot tell you who his client is.
- No one can name a single goal for the process improvement effort other than the achievement of a maturity level.
- There is a belief that no processes existed prior to the CMM initiative.
- Everyone feels like their only job is maturity level achievement.
- You are not allowed to use *any* of the old procedures, even ones that worked.
- Whenever you ask someone "why are you doing that?", they tell you that "the process requires it."
- People with software delivery responsibilities can recite CMM practices or the identification and titles of their organization's policies and procedures.
- Estimates for process overhead in development projects exceed 15% of the projects' total effort.
- The volume of standards and procedures increases, while the quantity and quality of delivered products decreases.
- People use words such as "audit," "inspection," and "compliance."
- People refer to "CMM" or "SEI" requirements.
- People make jokes about the "process police" or "process Gestapo."

What is the best way to promote software quality improvement without triggering a slash-and-burn frenzy?

Stelzer and Mellis [1998] provide an excellent set of suggestions of success factors learned from quality improvement initiatives in a large number of organizations that they study. Table 6.1 summarizes these findings.

TABLE 6.1

Organizational Change in Software Process Improvement

Successful factor of organizational change	Explanation
Change agents and opinion leaders	Change agents initiate and support the improvement projects at the corporate level, opinion leaders at a the local level.
Encouraging communication and collaboration	Communication efforts precede and accompany the improvement program (communication) and degree to which staff members from different teams and departments cooperate (collaboration).
Management commitment and support	Management at all organizational levels sponsor the change.
Managing the improvement project	Every process improvement initiative is effectively planned and controlled.
Providing enhanced understanding	Knowledge of current software processes and interrelated business activities is acquired and transferred throughout the organization.
Setting relevant and realistic objectives	Software processes are continually supported, maintained, and improved at a local level.
Stabilizing changed processes	Software processes are continually supported, maintained, and improved at a local level.
Staff involvement	Staff members participate in the improvement activities.
Tailoring improvement initiatives	Improvement efforts are adapted to the specific strengths and weaknesses of different teams and departments.
Unfreezing the organization	The "inner resistance" of an organizational system to change is overcome

Source: Stelzer, D. and Mellis, W., Success factors of organizational change in software process improvement, *Software Process — Improvement and Practice*, 4, 227–250, 1998.

6.3 Software Testing

What is the role of testing with respect to software quality?

Effective software testing will improve software quality. In fact, even poorly planned and executed testing will improve software quality if it finds defects. Testing is a life-cycle activity; testing activities begin from product inception and continue through delivery of the software and into maintenance. Collecting bug reports and assigning them for repair is also a testing activity. But as a life-cycle activity, the most valuable testing activities occur at the beginning of the project. Boehm and Basili [2005] report that finding and fixing a software problem after delivery is often 100 times more expensive than finding and fixing it during the requirements testing phase. And about 40 to 50% of the effort on current software projects is spent on avoidable rework.

Is there a difference between an error, a bug, a fault, and a failure?

There is more than a subtle difference between the terms error, bug, fault, and failure. Use of "bug" is, in fact, discouraged because it somehow implies that an error crept into the program through no one's action. The preferred term for an error in requirement, design, or code is "error" or "defect." The manifestation of a defect during the operation of the software system is called a fault. A fault that causes the software system to fail to meet one of its requirements is called a failure.*

Is there a difference between verification and validation?

Verification, or testing, determines whether the products of a given phase of the software development cycle fulfill the requirements established during the previous phase. Verification answers the question "Am I building the product right?"

Validation determines the correctness of the final program or software with respect to the user's needs and requirements. Validation answers the question "Am I building the right product?"

What is the purpose of testing?

Testing is the execution of a program or partial program with known inputs and outputs that are both predicted and observed for the purpose of finding faults or deviations from the requirements. Although testing will flush out errors, this is just one of its purposes. The other is to increase trust in the system. Perhaps once, software testing was thought of as intended to remove all errors. However, testing can only detect the presence of errors, not the absence of them; therefore, it can never be known when all errors have been detected. Instead, testing must increase faith in the system, even though it may still contain undetected faults, by ensuring that the software meets its requirements. This objective places emphasis on solid design techniques and a well-developed requirements document. Moreover, a formal test plan that provides criteria used in deciding whether the system has satisfied the requirements documents must be developed.

What is a good test?

A good test is one that has a high probability of finding an error. A successful test is one that uncovers an error.

* Some define a fault as an error found prior to system delivery and a defect as an error found after delivery.

What are the basic principles of software testing?

The following principles should always be followed [Pressman 2005]:

- All tests should be traceable to customer requirements.
- Tests should be planned long before testing begins.
- Remember that the Pareto principle applies to software testing.
- Testing should begin "in the small" and progress toward testing "in the large."
- Exhaustive testing is not practical.
- To be most effective, testing should be conducted by an independent party.

These are the most helpful and practical rules for the tester.

How do I start testing activities during the requirements engineering process?

Testing is a well-planned activity and should not be conducted willy nilly, nor undertaken at the last minute, just as the code is being integrated. The most important activity that the test engineer can conduct during requirements engineering is to ensure that each requirement is testable. A requirement that cannot be tested cannot be guaranteed and, therefore, must be reworked or eliminated. For example, a requirement that says "the system shall be reliable" is untestable. On the other hand, "the MTBF for the system shall be not less than 100 hours of operating time" may be a desirable level of reliability and can be tested; that is, demonstrated to have been satisfied.

There are other testing activities that are then conducted during design and coding.

What test activities occur during software design and code development?

During the design process, test engineers begin to design the corresponding test cases based on an appropriate methodology. The test engineers and design engineers work together to ensure that features have sufficient testability. Often, the test engineer can point out problems during the design phase, rather than uncover them during testing.

There are a wide range of testing techniques for unit testing, integration testing, and system level testing. Any one of these test techniques can be either insufficient or computationally unfeasible. Therefore, some combination of testing techniques is almost always employed.

What is unit level testing?

Several methods can be used to test individual modules or units. These techniques can be used by the unit author (sometimes called desk checking) and by the independent test team to exercise each unit in the system. These techniques can also be applied to subsystems (collections of modules related to the same function). The techniques to be discussed include black box and white box testing.

What is black box testing?

In black box testing, only inputs and outputs of the unit are considered; how the outputs are generated based on a particular set of inputs is ignored. Such a technique, being independent of the implementation of the module, can be applied to any number of modules with the same functionality.

Some widely used black box testing techniques include:

- exhaustive testing
- boundary value testing
- random test generation
- worst case testing

An important aspect of using black box testing techniques is that clearly defined interfaces to the modules are required. This places additional emphasis on the application of Parnas partitioning and interface segregation principles to module design.

What is exhaustive testing?

Brute force or exhaustive testing involves presenting each code unit with every possible input combination. Brute force testing can work well in the case of a small number of inputs each with a limited input range; for example, a code unit that evaluates a small number of Boolean inputs. However, a major problem with brute force testing is the combinatorial explosion in the number of test cases. For example, suppose a program takes as input five 32-bit numbers and outputs one 32-bit number. Taking a purely black box testing perspective, to exhaustively test we need to test all possible combinations of five 32-bit inputs. That is we have $2^{32} \cdot 2^{32} \cdot 2^{32} \cdot 2^{32} \cdot 2^{32} = 2^{160}$ test cases. Even if we could randomly generate and run these test cases, say, one every 1 μsec, the entire test set would take more than 4.6×10^{34} years to complete!

What is boundary value testing?

Boundary value or corner case testing solves the problem of combinatorial explosion by testing some very tiny subset of the input combinations identified as meaningful "boundaries" of input.

For example, consider the code previously discussed with five 32-bit inputs. If the test inputs are restricted to every combination of the min, max, and nominal values for each input, then the test set would consist of $3^5 = 243$ test cases. A test set of this size can be handled easily with automatic test case generation.

A stronger version of this testing could be found if we test values just less than and just greater than each of the boundaries, or we could select more than one nominal value for each input.

What is random test case generation?

Random test case generation, or statistically based testing, can be used for both unit and system level testing. This kind of testing involves subjecting the code unit to many randomly generated test cases over some period of time. The purpose of this approach is to simulate execution of the software under realistic conditions.

The randomly generated test cases are based on determining the underlying statistics of the expected inputs. The statistics are usually collected by expert users of similar systems or, if none exist, by informed guessing. The intent of this kind of testing is to simulate typical usage of the system.

The major drawback of such a technique is that the underlying probability distribution functions for the input variables may be unavailable or incorrect. Therefore, randomly generated test cases are likely to miss conditions with low probability of occurrence. These are precisely the kind of situations that may be overlooked in the design of the module. Failing to test these scenarios is an invitation to disaster.

What is equivalence class testing?

Equivalence class testing involves partitioning the space of possible test inputs to a code unit or group of code units into a set of representative inputs. I like the analogy of the crash test dummies that auto manufacturers use to ensure the safety of automobiles. Auto manufacturers don't have a crash test dummy representing every possible human being. Instead, they use a handful of representative dummies — small, average, and large adult males; small, average, and large adult females; pregnant female; toddler, etc. These categories represent the equivalence classes.

In the same way, we can partition input sets. For example, suppose we are testing a module with an input from a sensor that has an expected range of [-1000, 1000]. We can partition this interval in a number of ways. One way might be to consider all the values <1000 to be in one equivalence class, those in the range [-1000, 1000] in another equivalence class, and those values >10,000 to be a third equivalence class. Then we select a representative input from each of those classes, say -5000, 0, and 5000 and test these cases. We could also combine equivalence class testing with boundary value testing and test the following inputs: -5000, -1000, 0, 1000, and 5000. We can strengthen the testing further by testing around both sides of the boundary values.

Are there any disadvantages to black box testing?

One disadvantage is that it can bypass unreachable or dead code. In addition, it may not test all of the control paths in the module. In other words, black box testing only tests what is expected to happen, not what wasn't intended. White box or clear box testing techniques can be used to deal with this problem.

What is white box testing?

White box testing (sometimes called clear or glass box testing) seeks to test the structure of the underlying code. For this reason it is also called structural testing.

Whereas black box tests are data driven, white box tests are logic driven; that is, they are designed to exercise all paths in the code unit. For example, in the reject mechanism functionality of the baggage inspection system, all error paths would need to be tested including those pathological situations that deal with simultaneous and multiple failures.

White box testing also has the advantage that it can discover those code paths that cannot be executed. This unreachable code is undesirable because it is likely a sign that the logic is incorrect, it wastes code space memory, and it might inadvertently be executed in the case of corruption of the computer's program counter.

The following white box testing strategies will be discussed (this is not an exhaustive list of white box testing techniques, however):

- DD path testing
- DU path testing
- McCabe's basis path method
- Code inspections
- Formal program proving

What is DD path testing?

DD path testing, or decision-to-decision path testing is a form of white box testing based on the control structure of the program. In DD testing, a graph representation of the control structure of the program is used to generate test cases that traverse the graph, essentially from one decision branch (for example, `if-then` statement) to another in well-defined ways. Depending on the strength of the testing, different combinations of paths are tested. A detailed explanation of DD path testing can be found in Jorgensen [2002].

What is DU path testing?

DU (define-use) path testing is a data-driven white box testing technique that involves the construction of test cases that exercise all possible definition, change, or use of a variable through the software system. Suppose, for example, a variable "acceleration" is defined as a floating-point variable somewhere in the software system but is accessed or changed throughout. Test cases would then be constructed that involve the setting and changing of that variable, and then observing how those changes propagate throughout the system.

DU path testing is actually rather sophisticated because there is a hierarchy of paths involving whether the variable is observed (and, for example, some decision made upon its value) or changed. Interested readers are referred to Jorgensen [2002].

What is McCabe's basis path method?

We will discuss McCabe's metric to determine the complexity of code. However, McCabe's metric can also be used to determine a lower bound on the number of test cases needed to traverse all linearly independent paths in a unit of code. McCabe also provides a procedure for determining the linearly independent paths by traversing the program graph. This technique is called the basis path method [McCabe 1976].

The basis path method begins with the selection of a baseline path, which should correspond to some "ordinary" case of program execution along one of the programs. McCabe advises choosing a path with as many decision nodes as possible.

Next, the baseline path is retraced and, in turn, each decision is reversed; that is, when a node of outdegree of greater than two is reached, a different path must be taken. Continuing in this way until all possibilities are exhausted generates a set of paths representing the set of basis vectors. Then a clever construction is followed to force the program graph to look like a vector space by defining the notions of scalar multiplication and addition along code paths.

The technique is relatively simple, but it is best to consult an excellent reference on testing, such as Jorgensen [2002], for further details.

What are code inspections?

In code inspections, the author of some collection of software presents each line of code to a review group of peer software engineers. Code inspections can detect errors as well as discover ways for improving the implementation. This audit also provides an opportunity to enforce coding standards.

Inspections have been shown to be a very effective form of testing. According to Boehm and Basili [2005], peer code reviews catch 60% of the defects. But when the reviews are directed (meaning, the reviewers are asked to focus on specific issues), then 35% more defects are caught than in non-directed reviews.

What is formal program proving?

Formal program proving is a kind of white box testing using mathematical techniques in which the code is treated as a theorem and some form of calculus is used to prove that the program is correct. This form of verification requires a high level of training and is useful, generally, for only limited purposes because of the intensity of activity required.

What is system integration testing?

Integration testing involves testing of groups of components integrated to create a system or sub-system. The tests are derived from the system specification. The principle challenge in integration testing is locating the source of the error when a test fails. Incremental integration testing reduces this problem.

Once individual modules have been tested, then subsystems or the entire system need to be tested. In larger systems, the process can be broken down into a series of subsystem tests and then a test of the overall system.

If an error occurs during system-level testing, the error must be repaired. Ideally, every test case involving the changed module must be rerun and all previous system level tests must be passed in succession. The collection of system test cases is often called a system test suite.

What is incremental integration testing?

This is a strategy that partitions the system in some way to reduce the code tested. Incremental testing strategy includes:

- top-down testing
- bottom-up testing
- other kinds of system partitioning

In practice, most integration involves a combination of these strategies.

What is top-down testing?

This kind of testing starts with a high-level system and integrates from the top-down, replacing individual components by stubs (dummy programs) where appropriate. For example, in Figure 6.3 the nodes represent code units and the arcs represent some calling or invocation (if they are methods) sequence between those code units. The shaded areas represent the collection of code units to be tested in the appropriate test cases, organized as test sessions.

The testers work their way down from the "top" of the system, which is the main program if it is written in a procedural manner. If it is an object-oriented program, then it is trickier to identify the sequence of method invocation. Sequence diagrams or use cases might be helpful in drawing this diagram. We will discuss this situation shortly.

In any case, where code units are not tested, or not yet written, appropriate stubs need to be written. The stubs would have the appropriate parameter interface but do nothing except perhaps indicate somehow that they were successfully called or invoked.

What is bottom-up testing?

Bottom up testing is the reverse of top-down testing in which we integrate individual components in levels, from the bottom up, until the complete

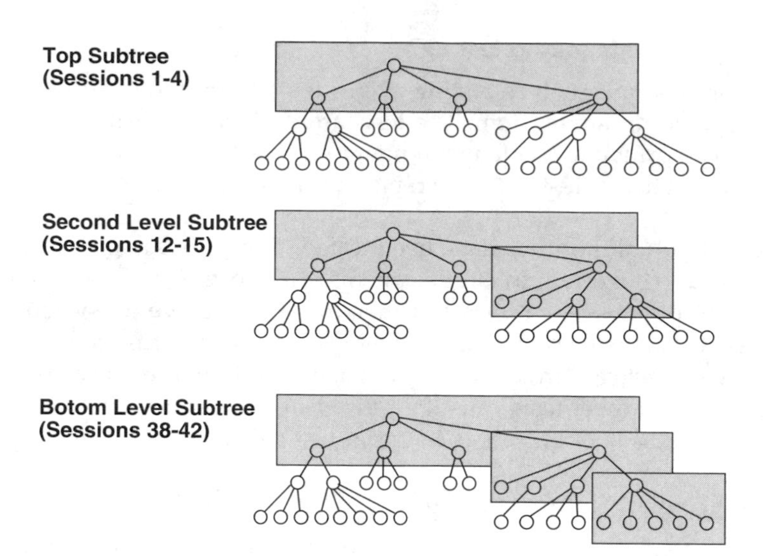

FIGURE 6.3
Top-down integration testing. (From Jorgensen, P.C., *Software Testing: A Craftsman's Approach*, 2nd ed., CRC Press, Boca Raton, FL, 2002. With permission.)

system is created. For example, in Figure 6.4, the testing sequence starts at the bottom level subtree and works its way to the top subtree. In bottom-up testing, test harnesses must be constructed; that is, skeleton code that represents the upper level functionality of the code must be created.

One of the disadvantages of bottom-up testing, however, is that integration errors are found later rather than earlier in the development of the system and, hence, systems-level design flaws that could require major reconstruction

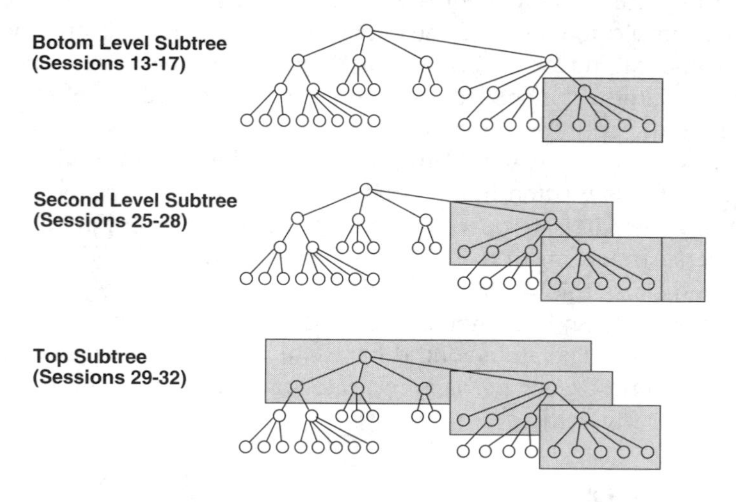

FIGURE 6.4
Bottom-up integration testing. (From Jorgensen, P.C., *Software Testing: A Craftsman's Approach*, 2nd ed., CRC Press, Boca Raton, FL, 2002. With permission.)

are found last. Finally, in this situation there is no visible working system until the last stage; therefore, it is harder to demonstrate progress to the clients.

What other kinds of system partitioning testing are there?

Other approaches attack related portions of the program in some fashion. These include pair-wise integration, sandwich integration, neighborhood integration testing, and interface testing. Interface testing describes a family of tests that deal with the specific integration issue of resolving problems that occur at the code unit interfaces. Object-oriented testing is also considered a systems integration testing strategy because other than testing the efficacy of class methods using unit-testing strategies, the behavior of objects as a whole requires the consideration of how they interact with other objects.

What is pair-wise integration testing?

This is a form of integration testing in which pairs of modules or functionality are tested together. Pair-wise integration testing is designed to reduce the effort needed to develop the stubs and drivers (the dummy code that plays the role of callees and callers in the program call graph) that are needed. Instead of waiting to write numerous stubs and drivers, the calling and called programs can be tested at the same time. Figure 6.5 illustrates the principle.

What is sandwich integration testing?

Sandwich integration is a combination of top-down and bottom up testing, which falls somewhere in between big-bang testing (one big test case) and testing of individual modules (see Figure 6.6). The technique tends to reduce stub and driver development costs, but is less effective in isolating errors.

What is neighborhood integration testing?

In this integration testing strategy, portions of program functionality that are logically connected somehow are tested in groups. As before, appropriate stubs and drivers are needed to deal with the interfaces to the code that is not being tested. (See Figure 6.7.)

What is interface testing?

This testing takes place when modules or subsystems are integrated to create larger systems. Therefore, it can take place during integration testing. The objective is to detect faults due to interface errors or invalid assumptions about interfaces. This kind of testing is particularly important for object-oriented development as objects are defined by their interfaces.

What kinds of interfaces can be tested?

Parameter interfaces can be tested to ensure that the correct data passed from one procedure to another. Shared memory interfaces can be tested by reading

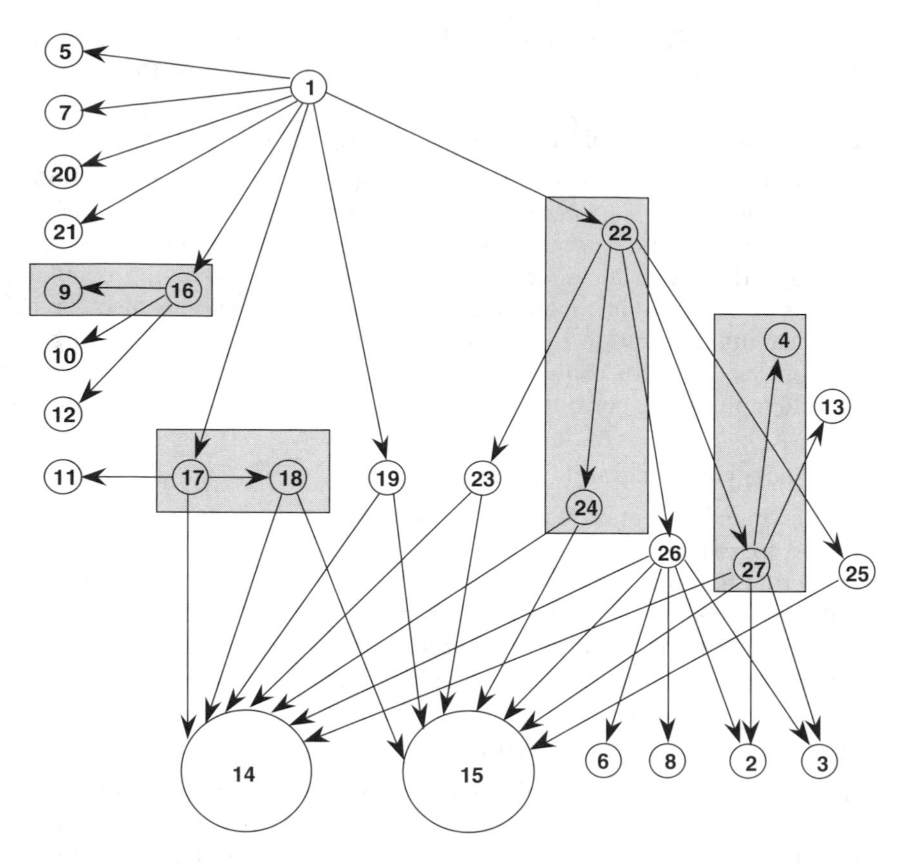

FIGURE 6.5
Some pair-wise integration sessions. The shaded pairs are tested together. (From Jorgensen, P.C., *Software Testing: A Craftsman's Approach,* 2nd ed., CRC Press, Boca Raton, FL, 2002. With permission.)

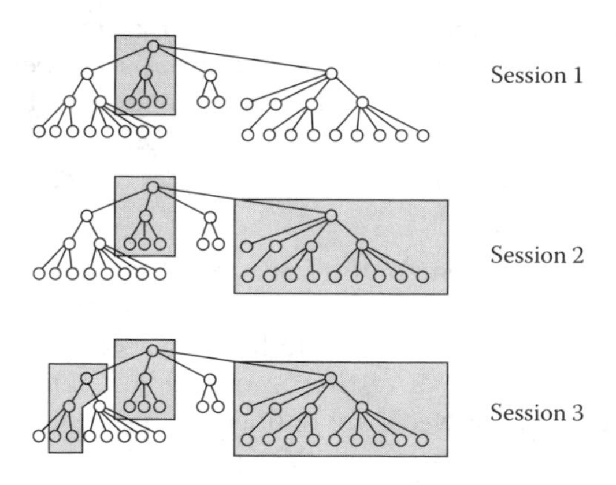

FIGURE 6.6
Some sandwich integration sessions. (From Jorgensen, P.C., *Software Testing: A Craftsman's Approach,* 2nd ed., CRC Press, Boca Raton, FL, 2002. With permission.)

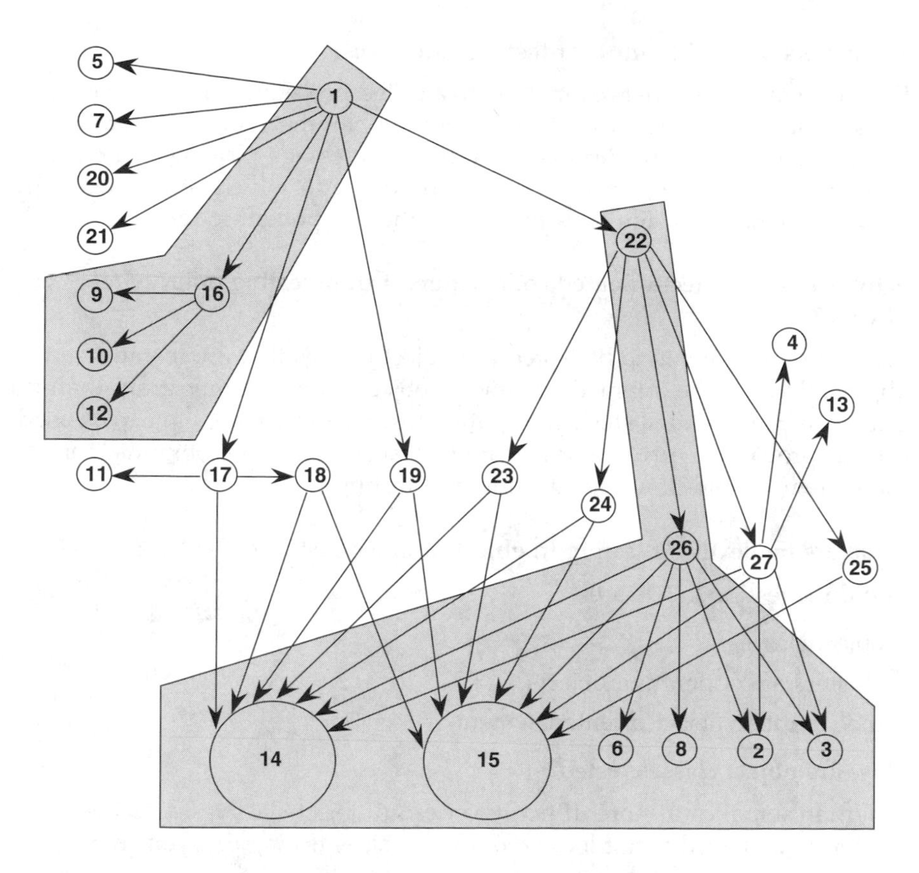

FIGURE 6.7
Two neighborhood integration sessions (shaded areas). (From Jorgensen, P.C., *Software Testing: A Craftsman's Approach,* 2nd ed., CRC Press, Boca Raton, FL, 2002. With permission.)

and writing global variables from all code units that can access those variables. Procedural interfaces can be tested to verify that each subsystem encapsulates a set of procedures to be called by other subsystems. Finally, message-passing interfaces can be tested to verify that subsystems correctly request services from other subsystems.

What kinds of errors can occur at the interfaces?

One error is interface misuse, where a calling component calls another component and makes an error in its use of its interface; for example, when parameters are in the wrong order. Another type of error is interface misunderstanding, which occurs when a calling component embeds incorrect assumptions about the behavior of the called component. Finally, timing errors can occur at the interfaces in that the called and the calling component operate at different speeds and out-of-date information is accessed.

What are some guidelines for testing interfaces?

First, design tests so that parameters to a called procedure are at the extreme ends of their ranges. Then, always test pointer parameters with null pointers. Use stress testing (to be discussed shortly) in message-passing systems and in shared-memory systems, and vary the order in which components are activated. Finally, design tests that cause the component to fail.

Why is testing object-oriented code different from testing other types of code?

First, the components to be tested are object classes that are instantiated as objects. Because the interactions among objects have a larger grain than do individual functions, systems integration testing approaches have to be used. But the problem is further complicated because there is no obvious "top" to the system for top-down integration and testing.

What are the levels of testing in object-oriented testing?

The three testing levels are:

object classes
clusters of cooperating objects
the complete object-oriented system

How are object classes tested?

Inheritance makes it more difficult to design object class tests, as the information to be tested is not localized. Object class testing can be achieved by

- testing all methods associated with an object
- setting and interrogating all object attributes
- exercising the object in all possible states

Note that methods can be tested using any of the black or white box testing techniques discussed for unit testing.

How can clusters of cooperating objects be tested?

Various techniques are used to identify clusters of objects using knowledge of the operation of objects and the system features that are implemented by these clusters. Cluster identification approaches include:

- use-case or scenario testing
- thread testing
- object interaction testing
- uses-based testing

Use-case or scenario testing is based on a user interaction with the system. This kind of cluster testing has the advantage that it tests system features as experienced by users.

Thread testing focuses on each thread. A thread is all of the classes needed to respond to a single external input. Each class is unit tested, and then the thread set is exercised.

Object interaction testing tests sequences of object interactions that stop when an object operation does not call on services from another object.

Uses-based testing begins by testing classes that use few or no server classes. It continues by testing classes that use the first group of classes, followed by classes that use the second group, and so on.

What is scenario testing?

This type of system testing for object-oriented systems involves identifying scenarios from use-cases and supplementing them with interaction diagrams that show the objects involved in the scenario.

What is worst case testing?

Worst case, or exception, testing involves those test scenarios that might be considered highly unusual and unlikely. Often these exceptional cases are exactly those for which the code is likely to be poorly designed and, therefore, to fail.

For example, in the baggage inspection system, while it might be highly unlikely that the system is to function at the maximum conveyor speed, this worst case still needs to be tested.

What is stress testing?

In this testing, the system is subjected to a large disturbance in the inputs; for example, baggage arriving at the maximum rate for an extended period. One objective of this kind of testing is to see how the system fails (gracefully or catastrophically).

Stress testing can also be useful in dealing with cases and conditions where the system is under heavy load; for example, when testing for memory or processor utilization in conjunction with other applications and operating system resources to determine if performance is acceptable.

Stress testing is particularly useful in distributed systems, which can exhibit severe degradation as a network becomes overloaded.

What is burn-in testing?

Burn-in testing is a type of system-level testing done in the factory, which seeks to flush out those failures appearing early in the life of the system, and thus to improve the reliability of the delivered product.

What is alpha testing?

This is system validation consisting of internal distribution and exercise of the software. This kind of testing is usually followed by beta testing.

What is beta testing?

This testing, which follows alpha testing, involves preliminary versions of validated software distributed to friendly customers who test the software under actual use. Based on feedback from the beta test sites, corrections or enhancements are added and then regression testing is performed.

What is regression testing?

Regression testing (which can also be performed at the unit level) is used to validate the updated software against the old set of test cases that have already been passed. Any new test case needed for the enhancements is then added to the test suite, and the software is validated as if it were a new product. Regression testing is also an integral part of integration testing as new modules are added to the tested subsystem.

What is cleanroom testing?

Cleanroom testing is more than a kind of system testing. It is really a testing philosophy. The principal tenant of cleanroom software development is that given sufficient time and care, error-free software can be written. Cleanroom software development relies heavily on code inspections and formal program validation. It is taken for granted that software specifications exist that are sufficient to completely describe the system.

The program code is developed by slowly "growing" features into the code, starting with some baseline of functionality. At each milestone an independent test team checks the code against a set of randomly generated test cases based on a set of statistics describing the frequency of use for each feature specified in the requirements.

Once a functional milestone has been reached, the development team adds to the "clean" code, using the same techniques as before. Thus, like an onion skin, new layers of functionality are added to the software system unit it has completely satisfied the requirements.

Successful projects have been developed in this way, in both academic and industrial environments. In any case, many of the tenants of cleanroom testing can be incorporated without completely embracing the methodology.

What is software fault injection?

Fault injection is a form of dynamic software testing that acts like "crash-testing" the software by demonstrating the consequences of incorrect code or data. Anyone who has ever tried to type a letter when the input called for a number is familiar with fault injection.

The main benefit of fault injection testing is that it can demonstrate that the software is unlikely to do what it shouldn't. Fault injection can also help to reveal new output states that before have never been observed or contemplated.

Fault injection can also be used as a test stoppage criterion; for example, test until fault injection no longer causes failure. Finally, it can be used as a safety case proposition — "Hey, we tested this system and injected all kinds of crazy faults and it didn't break" [Voas 1998].

When should you stop testing?

There are several criteria that can be used to determine when testing should cease. These include:

- When you run out of time.
- When continued testing causes no new failures.
- When continued testing reveals no new faults.
- When you can't think of any new test cases.
- When you reach a point of "diminishing returns."
- When mandated coverage has been attained.
- When all faults have been removed [Jorgensen 2002].

But the best way to know when testing is complete is when the test coverage metric requirements have been satisfied.

What are test coverage metrics?

Test coverage metrics are used to characterize the test plan and identify insufficient or excessive testing. Two simple metrics are requirements per test and tests per requirement. Obviously, every requirement should have one or more test associated with it. But every test, to be efficient, should test more than one requirement. However, one would not want an average of one test per requirement because this is essentially "big-bang" testing, and is ineffective in uncovering errors. Similarly, if a test covers too many requirements and it fails, it might be hard to localize the error.

Research is still ongoing to determine appropriate values for these statistics. But in any case, you can look for inconsistencies to determine if some requirements are not being tested thoroughly enough, if some requirements need more testing, and if some tests are too complex. Remember, if a test covers too many requirements and fails, it could be difficult to localize the error. There is always a trade-off between time and cost of testing vs. the comprehensiveness of testing.

How do I write a test plan?

The test plan should follow the requirements document item by item, providing criteria that are used to judge whether the required item has been met. A set of test cases is then written which is used to measure the criteria set out in the test plan. Writing such test cases can be extremely difficult when a user interface is part of the requirements.

The test plan includes criteria for testing the software on a module-by-module or unit level, and on a system or subsystem level; both should be incorporated in a good testing scheme. The system-level testing provides criteria for the hardware/software integration process. IEEE Standard 829–1998 (IEEE Standard for Software Test Documentation) can be helpful for those unfamiliar with software documents.

Are there automated tools for testing that can make the job easier?

Software testing can be an expensive proposition to conduct, but well worth the cost if done correctly. Testing workbenches provide a range of tools to reduce the time required and total testing costs. Most testing workbenches are open systems because testing needs are organization-specific.

What are some testing tools that I can use?

The xUnit frameworks developed by Kent Beck make object-oriented unit testing more accessible. xUnit (which stands for a family of testing frameworks, CUnit for C code, JUnit for Java code, PyUnit for Python, and so on.

The xUnit frameworks are particularly helpful in automating tests for regression testing.

6.4 Metrics

What are some motivations for measurement?

The key to controlling anything is measurement. Software is no different in this regard, but the question arises "what aspects of software can be measured?" Chapter 2 introduced several important software properties and alluded to their measurement. It is now appropriate to examine the measurement of these properties and show how this data can be used to monitor and manage the development of software.

Metrics can be used in software engineering in several ways. First, certain metrics can be used during software requirements development to assist in cost estimation. Another useful application for metrics is benchmarking. For example, if a company has a set of successful systems, then computing metrics for those systems yields a set of desirable and measurable characteristics with which to seek or compare in future systems.

Most metrics can also be used for testing in the sense of measuring the desirable properties of the software and setting limits on the bounds of those criteria. Or they can be used during the testing phase and for debugging purposes to help focus on likely sources of errors.

Of course, metrics can be used to track project progress. In fact, some companies reward employees based on the amount of software developed per day as measured by some of the metrics to be discussed (e.g., delivered source instructions, function points, or lines of code).

Finally, as Kelvin's quote at the start of the chapter suggests, a quality culture depends upon measurement. Such an approach is similar in other kinds of engineering.

So what kinds of things can we measure in software?

We can measure many things. Typical candidates include:

- lines of code
- code paths
- defect rates
- change rates
- elapsed project time
- budget expended

Is the lines of code metric useful?

The easiest characteristic of software that can be measured is the number of lines of finished source code. Measured as thousands of lines of code (KLOC), the "clock" metric is also referred to as delivered source instructions (DSI) or noncommented source code statements (NCSS). That is, we count executable program instructions (excluding comment statements, header files, formatting statements, macros, and anything that does not show up as executable code after compilation or cause allocation of memory).

Another related metric is source lines of code (SLOC), the major difference being that a single source line of code may span several lines. For example, an `if-then-else` statement would be a single SLOC, but many delivered source instructions.

While the "clock" metric essentially measures the weight of a printout of the source code, thinking in these terms makes it likely that the usefulness of KLOC will be unjustifiably dismissed as supercilious. But isn't it likely that 1000 lines of code is going to have more defects than 100 lines of code? Wouldn't it take longer to develop the latter than the former? Of course, the answer is dependent upon the complexity of the code.

What are the disadvantages of the LOC metric?

One of the main disadvantages of the using lines of source code as a metric is that it can only be measured after the code has been written. While lines of code can be estimated, this approach is far less accurate than measuring the code after it has been written.

Another criticism of the KLOC metric is that it does not take into account the complexity of the software involved. For example, 1000 lines of print statements probably does have less errors than 100 lines of a complex imaging algorithm.

Nevertheless, KLOC is a useful metric, and in many cases is better than measuring nothing. Many other metrics are fundamentally based on lines of code.

What is the delta lines of code metric?

Delta KLOC measures how many lines of code change over some period of time. Such a measure is useful, perhaps, in the sense that as a project nears the end of code development, Delta KLOC would be expected to be small. Other, more substantial, metrics are also derived from KLOC.

What is McCabe's metric?

To attempt to measure software complexity, McCabe [1976] introduced a metric, cyclomatic complexity, to measure program flow-of-control. This concept fits well with procedural programming but not necessarily with object-oriented programming, though there are adaptations for use with the latter. In any case, this metric has two primary uses:

- to indicate escalating complexity in a module as it is coded and therefore assisting the coders in determining the "size" of their modules
- to determine the upper bound on the number of tests that must be designed and executed

Use of the McCabe metric in calculating the minimum number of test cases needed to traverse all linearly independent code paths was already discussed.

How does McCabe's metric measure software complexity?

The cyclomatic complexity is based on determining the number of linearly independent paths in a program module; suggesting that the complexity increases with this number and reliability decreases.

To compute the metric, the following procedure is followed. Consider the flow graph of a program. Let e be the number of edges and n be the number of nodes. Form the cyclomatic complexity, C, as follows:

$$C = e - n + 2 \tag{6.1}$$

This is the most generally accepted form.*

Can you help me visualize the cyclomatic complexity?

To get a sense of the relationship between program flow and cyclomatic complexity, refer to Figure 6.8. Here, for example, a sequence of instructions has two nodes, one edge, and one region and, hence, would contribute a complexity of $C = 1 - 2 + 2 = 1$. This is intuitively pleasing as nothing could be less complex than a simple sequence.

On the other hand, the case statement, which has six edges, five nodes, and two regions, would contribute $C = 6 - 5 + 2 = 3$ to the overall complexity.

* There are other, equivalent, formulations; $C = e - n + p$ and $C = e - n + 2p$. The different forms arise from the transformation of an arbitrary directed graph to a strongly connected, directed graph obtained by adding one edge from the sink to the source node [Jorgensen 2002].

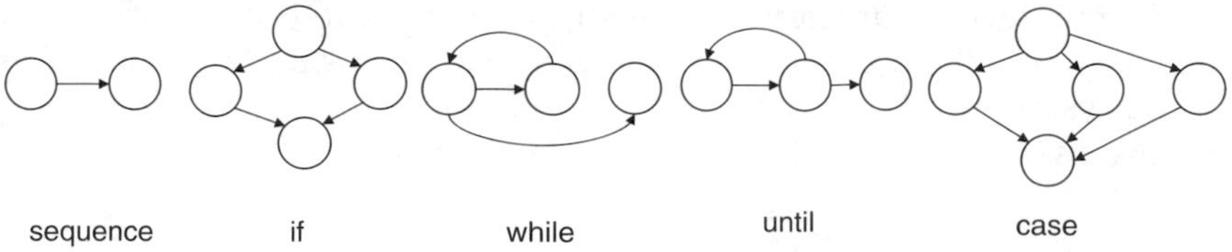

sequence if while until case

FIGURE 6.8
Correspondence of language statements and flow graph.

As a more substantial example, consider a segment of code extracted from the noise reduction portion of the baggage inspection system. The procedure calls between modules **a**, **b**, **c**, **d**, **e**, and **f** are depicted in Figure 6.9.

Here, then, $e = 9$ and $n = 6$ yielding a cyclomatic complexity of $C = 9 - 6 + 2 = 5$.

Can the computation of McCabe's metric be automated?

Computation of McCabe's metric can be done easily during compilation by analyzing the internal tree structure generated during the parsing phase (see Chapter 4). However, commercial tools are available to perform this analysis.

What are Halstead's metrics?

Halstead's metrics measure information content, or how intensively the programming language is used. Halstead's metrics are based on the number of distinct, syntactic elements and begin-end pairs (or their equivalent, such as open and closed curly braces in Java or C). From these a statistic for program length is determined. I will omit the equations because they are rarely computed by hand. From these statistics, a "program vocabulary," V,

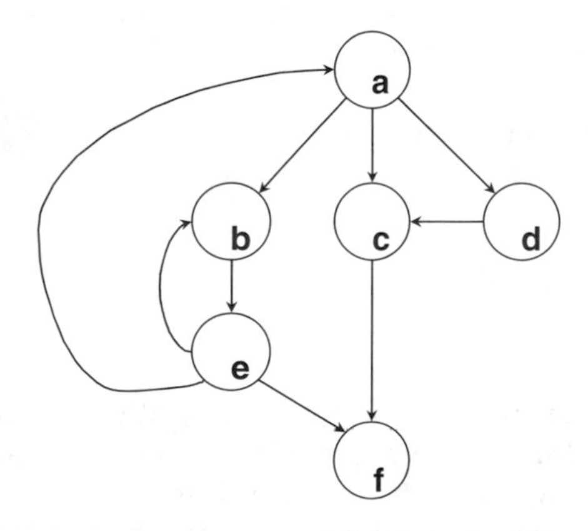

FIGURE 6.9
Flow graph for noise reduction code for the baggage inspection system.

and program level, L, are derived. L is supposed to be a measure of the level of abstraction of the program. It is believed that increasing this number will increase system reliability.

In any case, from V and L the effort, E, is defined as

$$E = V/L \tag{6.2}$$

Decreasing the effort level is believed to increase reliability as well as ease of implementation.

In principle, the program length can be estimated and, therefore, is useful in cost and schedule estimation. The length is also a measure of the "complexity" of the program in terms of language usage and, therefore, can be used to estimate defect rates.

Are Halstead's metrics still used?

Halstead's metrics, though dating back 30 years, are still widely used and tools are available to automate calculations.

What are function points?

Function points (FPs) were introduced in the late 1970s as an alternative to metrics based on simple source line count. The basis of FPs is that as more powerful programming languages are developed, the number of source lines necessary to perform a given function decreases. Paradoxically, however, the cost per LOC measure indicated a reduction in productivity, as the fixed costs of software production were largely unchanged.

The solution to this effort estimation paradox is to measure the functionality of software via the number of interfaces between modules and subsystems in programs or systems. A big advantage of the FP metric is that it can be calculated before any coding occurs.

What are the primary drivers for FPs?

The following five software characteristics for each module, subsystem, or system represent its FPs:

Number of inputs to the application (I)
Number of outputs (O)
Number of user inquiries (Q)
Number of files used (F)
Number of external interfaces (X)

In addition, the FP calculation takes into account weighting factors for each aspect that reflects their relative difficulty in implementation. These coefficients vary depending on the type of application system. Then complexity factor adjustments can be applied for different types of application domains. The full set of coefficients and corresponding questions can be found by consulting an appropriate text on software metrics.

TABLE 6.2

Programming Language and Lines of Code per FP

Language	LOC per FP
Assembly	320
C	128
Fortran	106
C++	64
Visual Basic	32
Smalltalk	22
SQL	12

Source: Adapted from [Jones 1998].

How do I interpret the FP value?

Intuitively, the higher FP, the more difficult the system is to implement. For the purposes of comparison, and as a management tool, FPs have been mapped to the relative lines of source code in particular programming languages. These are shown in Table 6.2.

For example, it seems intuitively pleasing that it would take many more lines of assembly language code to express functionality than it would in a high-level language like C. In the case of the baggage inspection system, with FP = 241, it might be expected that about 31,000 LOC would be needed to implement the functionality. In turn, it should take many less LOC to express that same functionality in a more abstract language such as C++. The same observations that apply to software production might also apply to maintenance as well as to the potential reliability of software.

How widely is the FP metric used?

The FP metric is widely used in business applications, but not nearly as much in embedded systems. However, there is increasing interest in the use of FPs in real-time embedded systems, especially in large-scale real-time databases, multimedia, and Internet support. These systems are data-driven and often behave like the large-scale transaction-based systems for which FPs were developed.

The International Function Point Users Group maintains a Web database of weighting factors and FP values for a variety of application domains. These can be used for comparison.

What are feature points?

Feature points are an extension of FPs developed by Software Productivity Research, Inc. in 1986. Feature points address the fact that the classical FP metric was developed for management information systems and, therefore, are not particularly applicable to many other systems, such as real-time, embedded, communications, and process control software. The motivation is that these systems exhibit high levels of algorithmic complexity, but sparse inputs and outputs.

The feature point metric is computed in a manner similar to the FP except that a new factor for the number of algorithms, A, is added.

Are there special metrics for object-oriented software?

While any of the previously discussed metrics can been used in object-oriented code, other metrics are better suited for this setting. Object-oriented metrics are computed on three levels:

methods
classes
packages

What are some method level metrics?

Often the lines of code, McCabe cyclomatic complexity, or even Halstead's metrics are used.

What are commonly used class level metrics?

The most widely used set of metrics for the object-oriented software, the Chidamber and Kemerer (C&K) metrics, contain metrics that are at the class level. These are primarily applied to the concepts of classes, coupling, and inheritance.

Some of the C&K metrics are

- Weighted Methods per Class (WMC) — the sum of the complexities of the methods (method complexity is measured by cyclomatic complexity, CC).
- Response for a Class (RFC) — the number of methods that can be invoked in response to a message to an object of the class or by some method in the class. Includes all methods accessible within the class hierarchy.
- Lack of Cohesion in Methods (LCOM) — the dissimilarity of methods in a class by instance variable or attributes.
- Coupling between Object Classes (CBO) — the number of other classes to which a class is coupled. Measured by counting the number of distinct non-inheritance related class hierarchies on which a class depends.
- Depth of Inheritance Tree (DIT) — the maximum number of steps from the class node to the root of the tree. Measured by the number of ancestor classes.
- Number of Children (NOC) — the number of immediate subclasses subordinate to a class in the hierarchy.

What are some package level metrics?

Martin's package cohesion metrics are based on the principles of good OOD discussed in Chapter 3.

- Afferent Coupling (A) — the number of other packages that depend upon classes within this package. This metric is an indicator of the package's responsibility.
- Efferent Coupling — the number of other packages that the classes in this package depend upon. This metric is an indicator of the package's independence.
- Abstractness — the ratio of the number of abstract classes (and interfaces) in this package to the total number of classes in the analyzed package.
- Instability (I) — the ratio of efferent coupling to total coupling.
- Distance from the Main Sequence — the perpendicular distance of a package from the idealized line, A + I = 1.
- Law of Demeter — this can be informally stated as "only talk to your friends," that is, limit the intellectual distance from any object and anything it uses. This metric is measured by the number of methods called subsequently by the external objects participating in the method chain.

Are there other kinds of object-oriented metrics?

Yes, there seem to be more metrics for object-oriented code than for procedural code. Other object-oriented metrics track things like component dependencies or rely on various features (for example, depth, cycles, or fullness) of graphs depicting packages or classes relationships.

What are object points?

This is a function point-like metric and is not specifically intended for use with object-oriented code, despite its name. Like FPs, it is a weighted estimate of the following visible program features:

- number of separate screens
- number of reports
- number of third generation language modules needed to support fourth generation language code

Object points are an alternative to FPs when fourth generation languages are used.

What are use case points?

Use case points (UCPs) allow the estimation of an application's size and effort from its use cases. UCPs are based on the number of actors, scenarios, and various technical and environmental factors in the use case diagram. The UCP equation is based on four variables:

technical complexity factor (TCF)
environment complexity factor (ECF)
unadjusted use case points (UUCP)
productivity factor (PF)

which yield the equation:

$$UCP = TCP * ECF * UUCP * PF \quad (6.3)$$

UCPs are a relatively new estimation technique.

This is all so confusing; which metrics should I use?

The goal question metric (GQM) paradigm is a helpful framework for selecting the appropriate metrics.

What is the GQM technique?

GQM is an analysis technique that helps in the selection of an appropriate metric. To use the technique, you follow three simple rules. First, state the goals of the measurement; that is, what is the organization trying to achieve? Next, derive from each goal the questions that must be answered to determine if the goals are being met. Finally, decide what must be measured in order to be able to answer the questions.

Can you give a simple example?

Suppose one of your organization's goals is to evaluate the effectiveness of the coding standard. Some of the associated questions you might ask to assess if this goal has been achieved are:

Who is using the standard?
What is coder productivity?
What is the code quality?

For these questions, the corresponding metrics might be:

What proportion of coders are using the standard?

How have the number of LOC or FPs generated per day per coder changed?

How have appropriate measures of quality for the code changed? (Appropriate measures might be errors found per LOC or cyclomatic complexity.)

Now that this framework has been established, appropriate steps can be taken to collect, analyze, and disseminate data for the study.

What are some objections to using metrics?

Some widely used objections are that metrics can be misused or that they are a costly and unnecessary distraction. For example, metrics related to the number of LOC imply that the more powerful the language, the less productive the programmer. Hence, obsessing with code production based on LOC is a meaningless endeavor.

Another objection is that the measuring of the correlation effects of a metric without clearly understanding the causality is unscientific and dangerous. For example, while there are numerous studies suggesting that lowering the cyclomatic complexity leads to more reliable software, there just isn't any real way to know why. Obviously, the arguments about the complexity of well-written code vs. "spaghetti code" apply, but there is no way to show the causal relationship. So, the opponents of metrics might argue that if a study of several companies shows that software written by software engineers who always wore yellow shirts had statistically significant less defects in their code, then companies would start requiring a dress code of yellow shirts! This illustration is, of course, hyperbole, but the point of correlation versus causality is made.

While in many cases these objections might be valid, like most things metrics can be either useful or harmful depending on how they are used (or abused).

6.5 Fault Tolerance

What are checkpoints?

Checkpoints are a way to increase fault tolerance. In this scheme, intermediate results are written to memory at fixed locations in code, called checkpoints, for diagnostic purposes (Figure 6.10). The data in these locations can be used for debugging purposes, during system verification, and during system operation verification.

If the checkpoints are used only during testing, then this code is known as a test probe. Test probes can introduce subtle timing errors, which are discussed later.

What are recovery blocks?

Fault tolerance can be further increased by using checkpoints in conjunction with predetermined reset points in software. These reset points mark recovery blocks in the software.

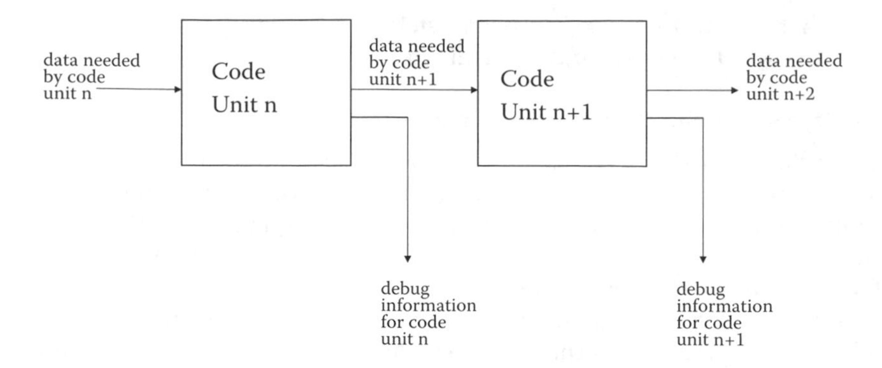

FIGURE 6.10
Checkpoint implementation.

At the end of each recovery block, the checkpoints are tested for "reasonableness." If the results are not reasonable, then processing resumes at the beginning of that recovery block or at some point in the previous one (see Figure 6.11).

The point, of course, is that some hardware device (or another process that is independent of the one in question) has provided faulty inputs to the block. By repeating the processing in the block, with presumably valid data, the error will not be repeated.

Each recovery block represents a redundant parallel process to the block being tested. Unfortunately, although this strategy increases system reliability, it can have a severe impact on performance because of the overhead added by the checkpoint and repetition of the processing in a block

What are software black boxes?

The software black box is used in certain mission critical systems. The objective of a software black box is to recreate the sequence of events that led to the software failure for the purpose of identifying the faulty code. The software

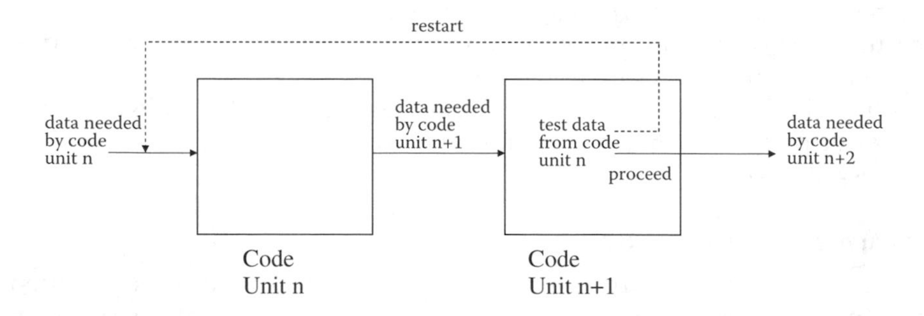

FIGURE 6.11
Recovery block implementation.

black box recorder is essentially a checkpoint that stores behavioral data during program execution.

As procedural code is executed, control of execution passes from one code module to the next. This occurrence is called a transition. These transitions can be represented by an $N \times N$ matrix, where N represents the number of modules in the system.

When each module i calls module j, the corresponding entry in the transition frequency matrix is incremented. From this process, a transition probability matrix is derived that records the likelihood that a transition will occur. This matrix is stored in nonvolatile memory.

Recovery begins after the system has failed and the software black box information is recovered. The software black box decoder generates possible functional scenarios from the transition probability matrix, allowing software engineers to reconstruct the most likely sequence of module execution that led to the failure.

What is N-version programming?

Sometimes a system can enter a state whereby it is rendered ineffective or deadlocked. This situation can be due to an untested flow-of-control in the software for which there is no escape or resource contention. For life-support systems, avionics systems, power plant control systems, and other types of systems, the results can be catastrophic.

In order to reduce the likelihood of this sort of catastrophic error, redundant processors are added to the system. These processors are coded to the same specifications but by different programming teams. It is, therefore, highly unlikely that more than one of the systems can lock up under the same circumstances. This technique is called N-version programming.

What is built-in-test software?

Built-in-test software is any hardware diagnostic software that is executed in real-time by the operating system. Built-in-test software can enhance fault tolerance by providing ongoing diagnostics of the underlying hardware.

Built-in-test software is especially important in embedded systems. For example, if the built-in-test software determines that a sensor is malfunctioning, then the software may be able to shut off the sensor and continue operation using backup hardware.

How should built-in-test software include CPU testing?

It is probably more important that the health of the CPU be checked than any other component of the system. A set of carefully constructed tests can be performed to test the health of the CPU circuitry.

Should built-in-test software test memory?

All types of memory, including nonvolatile memory, can be corrupted via electrostatic discharge, power surging, vibration, or other means. This damage can manifest itself either as a permutation of data stored in memory cells or as permanent damage to the cell itself. Damage to the contents of memory is called soft error, whereas damage to the cell itself is called hard error. All memory should be tested both at initialization and during normal processing, if possible. There are various algorithms for efficiently testing memory, which are described in embedded software design books such as *Software Engineering for Image Processing Systems* [Laplante 2004].

What about testing other devices?

Devices such as sensors, motors, cameras, and the like need to be tested continually or their own self-testing needs to be monitored by the software and appropriate action taken if the device falls.

6.6 Maintenance and Reusability

What is meant by software maintenance?

Software maintenance is the "...correction of errors, and implementation of modifications needed to allow an existing system to perform new tasks, and to perform old ones under new conditions..." [Dvorak 1994].

What is reverse engineering?

Generically, reverse engineering is the process of analyzing a subject system to identify its components. Reverse engineering is sometimes called renovation or reclamation. While there are negative connotations to reverse engineering as in theft of a design, reverse engineering, in some form, is essential for the improvement of the design or implementation or for recovery of documentation in the case of a system that may have been acquired legitimately from a third party.

What is an appropriate model for software reengineering?

Many embedded and engineering software systems are legacy systems; that is, they constitute the next generation of an existing system. Others borrow code from related systems. In any case, most systems need to have a long shelf life so that development costs can be recouped. Maintaining a system over a long period usually requires some form of reengineering; that is, a reverse flow through the software life cycle.

Figure 6.12 is a graphical representation of a reengineering process. The forward engineering flow represents a simple, three-phase waterfall model — requirements, design, and implementation.

Documentation recovery or redocumentation is the creation or revision of documentation from a given system, including requirements and design

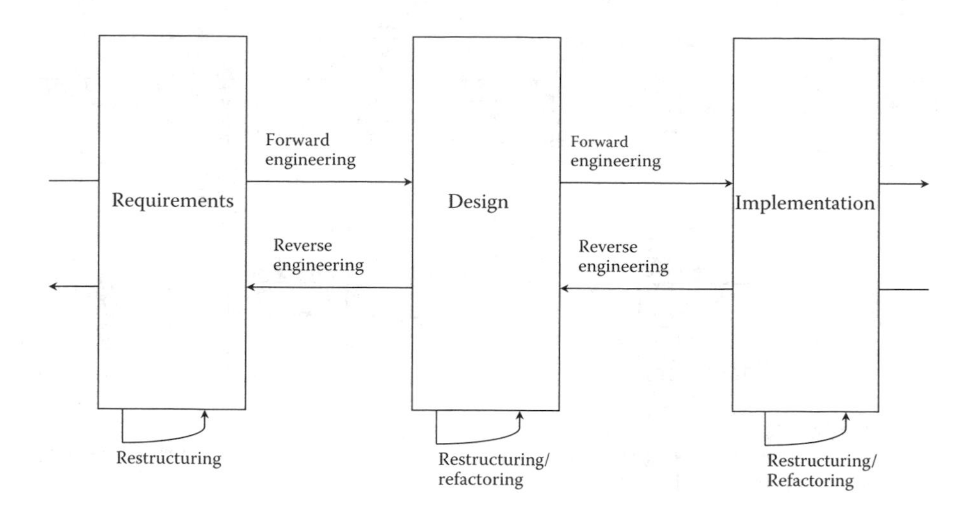

FIGURE 6.12
A reverse engineering process model.

documentation discovery. The need for redocumentation arises when there is poor or missing documentation for any of a number of reasons.

Design recovery is a subset of reverse engineering that recreates the design from code, existing documentation, personal insight, interviews with developers, and general knowledge. Again, the need for this arises in the case of poorly documented design, missing documentation, acquisition of a product from a company with inferior software engineering practices, and so on.

Restructuring is the transformation of one representation to another. In the case of code refactoring, the code is transformed so that the behavior is preserved. In design refactoring, the design is reengineered.

Since you like models so much, can you give me a maintenance process model?

Of the all of the phases, perhaps the maintenance model is the least understood. The maintenance phase generally consists of a series of reengineering processes to prolong the life of the system. There are three types of maintenance:

Adaptive — changes that result from external changes to which the system must respond.

Corrective — changes that involve maintenance to correct errors.

Perfective — all other maintenance including enhancements, documentation changes, efficiency improvements, and so on.

A widely adopted maintenance model illustrates the relationship between these various forms of maintenance (Figure 6.13).

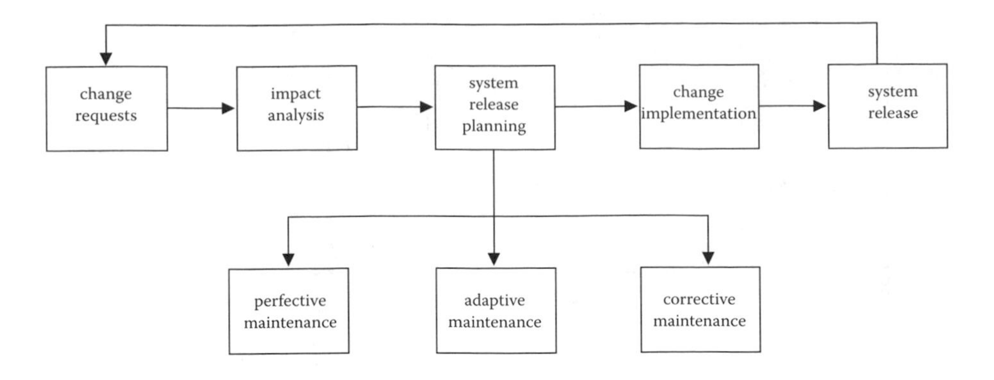

FIGURE 6.13
A maintenance process model. (Adapted from Sommerville, I., *Software Engineering*, 6th ed., Addison-Wesley, Boston, MA, 2000.)

The model starts with the generation of change requests by the customer, management, or the engineering team. An impact analysis is performed. This analysis involves studying the affect of the change on related components and systems and includes the development of a budget.

What is system release planning?

System release planning involves determination of whether the change is perfective, adaptive, or corrective. The nature of the change is crucial in determining whether the release needs to be made to all customers or specific customers, whether the release is going to be immediate or included in the next version, and so on. Finally, the change is implemented (invoking a mini-software life-cycle process from concept to acceptance testing), followed by the official release of the new version.

What is software reuse?

Pure software reuse is a highly sought prize in software engineering. It is clearly desirable to have a collection of mix-and-match, validated software components that could easily be pulled off the shelf for customized software applications. However, software reuse is virtually an exploitation of hard-learned experience. Even if software modules are not being explicitly reused, the lessons learned from previous but similar software projects should be carried forward.

Most of the cost savings can then be expected by reusing domain-specific models. To reuse domain-specific logic, however, developers must clearly separate domain logic from that of the application. They must also clearly distinguish domain-independent logic.

Therefore, the best way to begin a program of software reuse is to start small and learn by doing. Try to identify several small software modules that are good candidates for reuse and focus on preparing these modules for that reuse.

Are there special techniques for achieving reuse in procedural languages?

One technique that has been used in building program libraries involves domain analysis. Domain analysis views software code as functions with an input domain and output range based on the range of their inputs.

The approach is as follows. In a set-theoretic way, define the input and output domains for each module to be added to the program library. Then, determine the input/output dependencies between each module in the library and any candidate module to be added to the library. The existence of such dependencies determines the compatibility of the candidate module with the existing library modules. Of course, it is assumed that each candidate module has been validated, and is fully tested, at the module level.

For example, consider a program library that consists of trusted code unit A. Code unit A has an input domain of A_I and an output range of A_o. Now consider a new candidate code unit B, which has already been unit tested. Code unit B has an input domain of B_I and an output range of B_o. "Domains" and "ranges" mean the set of input and output variables of these modules and their ranges.

Now, if the output range of A (the variables that A could change) does not intersect with the input range of B and vice versa, then module B may be added to the program library without further interdependence or compatibility testing. If the input range of B and the output range of A overlap, then interdependencies and compatibility need to be tested before adding A to the library. Formally,

$$\text{If } A_O \cap B_I = \phi \quad \text{and} \quad B_O \cap A_I = \phi \text{ then add A to the library}$$

$$\text{Else test further before adding} \tag{6.4}$$

As additional modules or code are added to the library, interdependence testing must be completed for all modules in the library. For example, if A and B are trusted software in the library and module C is a candidate for the library, it must now be tested against A and B before adding it. Formally,

$$\text{If } A_O \cap C_I = \phi \quad \text{and} \quad B_O \cap C_I = \phi \quad \text{and}$$

$$C_O \cap B_I = \phi \quad \text{and} \quad B_O \cap C_I = \phi \quad \text{then add A to the library}$$

$$\text{Else test further before adding} \tag{6.5}$$

It is easy to see that the level of effort grows rapidly as new code is added to the trusted program library.

Are there special techniques for achieving reuse in object-oriented languages?

In object-oriented systems, use of design patterns following the OOD principles previously described promotes reusability. For example, one way is to employ the protected variation (Parnas partitioning) principle by identifying those design aspects that are likely to change and build a stable interface around them.

Design patterns can loosen the binding between program components enabling certain types of program evolution to occur with minimal changes to the program itself. However, to make good use of design patterns, the application's design process must undergo at least two iterations over the project life cycle.

When is it appropriate not to reuse software?

It is sometime desirable to plan not to reuse certain code. For example, throw-away prototypes are intentionally not to be reused. In other cases, it may not be desirable to try to reuse code that is of limited value. For example, a set of utilities intended for very specific hardware or that serves a very specialized function is probably not worth engineering for reuse when the hardware changes or becomes obsolete.

In any event, reuse of code that was not designed and coded for reuse can create many problems. For example, when a "quick-and-dirty" program becomes a widely used tool, it can present a maintenance nightmare.

What is Pareto's principle?

Pareto was a late 19th and early 20th century Italian mathematician and economist who was interested in the laws of chance. His observations can be applied in several ways to software reuse and engineering. For example, Pareto's principle might suggest that

- 20% of the code contributes to 80% of the cost of software development.
- 20% of the code contributes 80% of the errors.
- 20% of the errors account for 80% of the cost to fix.
- 20% of the modules consume 80% of the execution time.

The percentages are, of course, arbitrary. But these observations provide insight into how to approach software reuse, testing, and effort planning. For example, it would be helpful to identify the 20% of software that is the most expensive to develop and plan to reuse that software. The other 80% that is relatively easy to develop might not be a prime candidate for reuse. Checkpoints and software black boxes can help to collect code unit execution frequency to identify the high-use code.

What is the "Second System Effect?"

The Second System Effect first characterized by Brooks [1995] explains why software maintenance for legacy systems presents such challenges. This phenomenon is discussed in *The Mythical Man Month*, a series of essays on software project management by Brooks published in the early 1980s. The essays are relevant even today. Brooks notes that "second systems" or the next generation of a delivered system tend to be over-engineered. That is, there is a tendency to carry over and refine techniques whose existence has been made obsolete by changes in basic system assumptions. Doing so tends to make these systems hard to maintain, unwieldy, and unreliable.

Consider, for example, a baggage handling system that was developed in the 1970s for hardware that is no longer available. In a second system, the underlying hardware may have been modernized. Hence, carrying over old design decisions can be disastrous. Embedded systems tend to be based on carry-over software, often originally written in Fortran, C, or assembly language and even BASIC. In some cases, C code is simply "objectified" by wrapping the C code in such a way that it can be compiled as C++ code.

Brooks suggests that the way to avoid this effect is to insist on a project leader who has had experience with at least two systems. In this recommendation, Brooks recognizes that software houses tend to assign new software engineers to maintain old legacy systems, while the more senior engineers are assigned to new software development. While new projects may be more glamorous, younger engineers may not have the confidence or experience to challenge bad design decisions on a legacy system. Hence, it is probably better to have a combination of experience and youth assigned to both new and legacy system software development.

6.7 Further Reading

Basili, V.R., Caldiera, G., and Rombach, H.D., The goal question metric approach, in Boehm, B., Software risk management: principles and practice, *IEEE Software*, 8(1), 32–41, 1991.

Boehm, B. and Basili, V., Software defect reduction top-10 list, in *Foundations of Empirical Software Engineering*, Boehm, B., Rombach, H.D., and Zelkowitz, M., Eds., Springer, Secaucus, NJ, 2005, 427–431.

Brooks, F.W., *The Mythical Man Month, 20th Anniversary Edition*, Addison-Wesley, Boston, MA, 1995.

Chidamber, S.R. and Kemerer, C.F., A metrics suite for object oriented design, *IEEE Trans. Software Eng.*, 20(6), 476–493, 1994.

Darcy, D. and Kemerer, C., OO metrics in practice, *IEEE Software*, 22(6), 17–19, 2005.

Dvorak, J., Conceptual entropy and its effect on class hierarchies, *Computer*, 27(6), 59–63, 1994.

Eickelmann, N., Measuring maturity goes beyond process, *IEEE Software*, 12–13, 2004.

Godfrey, M. and Tu, Q. Growth, evolution and structural change in open source software, *Proc. 2001 Intl. Workshop on Principles of Software Evolution* (IWPSE-01), Vienna, September 2001.

Grady, R., *Practical Software Metrics for Project Management & Process Improvement*, Prentice Hall, Englewood Cliffs, NJ, 1992.

Hardgrave, W. and Armstrong, N.Y, Software process improvement: it's a journey not a destination, *Commun. ACM*, 49(11), 93–96, 2005.
Jones, C., *Estimating Software Costs,* McGraw-Hill, New York, 1998.
Jorgensen, P.C., *Software Testing: A Craftsman's Approach,* 2nd ed., CRC Press, Boca Raton, FL, 2002.
Kandt, K., *Software Engineering Quality Practices*, Auerbach Publications, Boca Raton, FL, 2005.
Koch, S., Evolution of open source software systems — A large-scale investigation, in *Proc. 1st Intl. Conf. OSS*, Scotto, M. and Succi, G., Eds., July 11–15, 2005, Genova, 148–153.
Lakos, J., *Large-Scale C++ Ssoftware Design*, Addison-Wesley, Boston, MA, 1996.
Lange, C.F.J., Chaudron, M.R.V., and Muskens, J., UML software architecture and design description, *IEEE Software*, 23(2) 40–46, 2006.
Laplante, P.A., *Software Engineering for Image Processing Systems*, CRC Press, Boca Raton, FL, 2004.
Machiavelli, N., *The Prince*, 1513.
McCabe, T., A software complexity measure, *IEEE Trans. Software Eng.*, SE-2, 308–320, 1976.
Martin, R.C., *Agile Software Development: Principles, Patterns, and Practices,* Prentice-Hall, Englewood Cliffs, NJ, 2002.
Munson, J., *Software Engineering Measurement*, Auerbach Publications, Boca Raton, FL, 2003.
Nakakoji, K., Yamamoto, Y., Nishinaka, Y., Kishida, K., and Ye, Y., Evolution patterns of open-source software systems and communities, *Proc. Intl. Workshop on Principles of Software Evolution*, ACM Press, Orlando, FL, May 2002.
NIST, Planning Report 02-3, The Economic Impact of Inadequate Infrastructure for Software Testing, www.nist.gov/director/prog-ofc/report02-3.pdf.
Paulk, M.C. et al., *The Capability Maturity Model, Guidelines for Improving the Software Process*, Addison-Wesley, Boston, MA, 1995.
Pfleeger, S. L. Measuring software reliability, *IEEE Spectrum*, 55–60, 1992.
Pressman, R.S., *Software Engineering: A Practitioner's Approach*, 6th ed., McGraw-Hill, New York, 2005.
Raymond, E.S. The cathedral and the bazaar. musings on linux and open source by an accidental revolutionary, 2001, O'Reilly Media, Cambridge, MA.
Scacchi, W., Understanding open source software evolution, in *Software Evolution*, Madhavji, N.H., Lehman, M.M., Ramil, J.F., and Perry, D., Eds., John Wiley & Sons, New York, 2004.
Sharp, J. et al., Tensions around the adoption and evolution of software quality management systems: a discourse analytic approach, *J. Human-Computer Studies*, 61, 219–236, 2005.
Sommerville, I., *Software Engineering*, 6th ed., Addison-Wesley, Boston, MA, 2000.
Stelzer , D. and Mellis, W., Success factors of organizational change in software process improvement, *Software Process — Improvement and Practice*, 4, 227–250, 1998.
Voas, J.M. and McGraw, G., *Software Fault Injection: Inoculating Programs Against Errors*, John Wiley & Sons, New York, 1998.
Voas, J. and Agresti, W.W., Software quality from a behavioral perspective, *IT Pro*, 6(4), 46–50, 2004.
West, M., *Real Process Improvement Using CMMI*, Auerbach Publications, Boca Raton, FL, 2004.
Zubrow, D., Current trends in the adoption of the CMMI® product suite, *compsac, 27th Annu. Intl. Comp. Software Appl. Conf.*, 2003, 126–129.

7

Managing Software Projects and Software Engineers

Outline

- Software engineers are people too
- Project management basics
- Tracking and reporting progress
- Software cost estimation
- Project cost justification
- Risk management

7.1 Introduction

For some reason a stereotype exists that engineers lack people skills. In my experience, this unfair perception of engineers is prevalent in those who have little or no education in the sciences or mathematics. These folks apparently think that engineers, mathematicians, and scientists lack soft skills because they have mastered the more analytical disciplines.

Of course, there are engineers or scientists who have unpleasant personalities. But there are nonengineers with rotten demeanors, too. I have never seen a study by any behavioral scientists demonstrating that lousy personalities are found at a higher frequency in engineers than in any other profession. The point here is that it doesn't matter who you are or where you came from — if you are a manager of any type, even a software project manager, you need the right attitude, education, and experience to be a good software project manager.

In this chapter, we will examine the people aspects of software project management as well as the more technical issues of cost determination, cost justification, progress tracking and reporting, and risk management.

While some of the topics are unique to software projects, most of the ideas presented in this chapter can be applied across a wide spectrum of technical and nontechnical projects. Many of the discussions are adapted from my own book on antipatterns [Laplante and Neill 2006], my text on software engineering for imaging engineers [Laplante 2003], and various lectures that I have given on the subject to a variety of project engineers at many companies. The works of Raffo and colleagues [1999] and Morgan [2005] influence the discussion on cost justification.

7.2 Software Engineers Are People Too

What personnel management skills does the software project manager need?

A project manager needs to have an appropriate set of people skills and relevant technical skills. The people skills include team building, negotiation techniques, understanding of psychology and group dynamics, good motivational skills, and excellent communication skills (especially listening). Most of the people skills involve self-improvement, and it is beyond the scope of this book to delve too deeply into that. There are many good books on the self-improvement aspects of people management, some of which can be found in the Further Reading section of this chapter. I will discuss some of the team building aspects of people management, however.

But what's the big deal with "people issues"?

It is well known that the success of a project is directly related to the quality of talent employed and, more importantly, the manner in which the talent is deployed on the project. But too frequently project managers* view themselves as technical managers only, forgetting that human nature enters into technical situations.

Moreover, the special challenges of developing software are imposed on top of the already daunting challenge of managing human teams. Some people might consider the aspect of human resource management insignificant if the project team has enough technical skill. This is generally not true.

How does team chemistry involve software projects?

The key problem in most cases is that the chemistry of the team makes it impossible for the manager to overcome other constraints, such as technical, time, and budget, even with good people. Table 7.1 illustrates the four possible cases of good/bad management and good/bad team chemistry.

* The term manager as a general term for anyone who is the responsible charge for one or more other persons developing, managing, installing, supporting, or maintaining systems and software. Other typical titles include "Software Project Manager," "Technical Lead," and "Senior Developer."

TABLE 7.1

Four Possible Combinations of Good/Bad Management and Good/Bad Team Chemistry

	Good Team chemistry	Bad Team Chemistry
Good management	Likely success	Possible success
Bad management	Unlikely Success	Unlikely Success

In the case where both management and chemistry are good, the likelihood of project success (which itself, must be carefully defined) is high. In the case of bad management, success is unlikely even with good team chemistry because bad management will eventually erode morale. But when team chemistry is bad, good management can possibly lead to success.

Why is team chemistry so hard to manage?

One reason is that the number of working relationships grows as a polynomial function of n, the number of people on the team. This might be whimsically referred to as the "n-body problem." In fact, it can easily be shown that for n people on a team, there are $\frac{n(n-1)}{2}$ possible working relationships, any of which can sour. Furthermore, a working relationship is not transitive. So, for example, Roger may work well with Mary and Mary with Sue, but Roger and Sue may not work well together. Finally, complicating these interactions is intercultural differences and outsourcing of project components. All of these aspects must be considered when building and managing teams, planning projects, and dealing with difficult personnel situations. Too many cooks spoil the broth, or to paraphrase Brooks [1995], "adding manpower to a late software project makes it later."

7.2.1 Management Styles

What are some styles for leading teams?

There are almost as many management styles as there are people. But, traditionally, a small collection of paradigms can be used to more or less describe the management style of an individual or organization. Understanding these basic approaches can be helpful in understanding the motivations of customers, supervisors, and subordinates.

What is Theory X?

Theory X, perhaps the oldest management style, is closely related to the hierarchical, command-and-control model used by military organizations. Accordingly, this approach is necessary because most people inherently dislike work and will avoid it if they can. Hence, managers should coerce, control, direct, and threaten their workers in order to get the most out of them. A typical statement by a "Theory X manager" is "people only do what you audit."

What is Theory Y?

As opposed to Theory X, Theory Y holds that work is a natural and desirable activity. Hence, external control and threats are not needed to guide the organization. In fact, the level of commitment is based on the clarity and desirability of the goals set for the group. Theory Y posits that most individuals actually seek responsibility and do not shirk it, as Theory X proposes.

A Theory Y manager simply needs to provide the resources, articulate the goals, and leave the team alone. This approach doesn't always work, of course, because some individuals do need more supervision than others.

What is Theory Z?

Theory Z is based on the philosophy that employees will stay for life with a single employer when there is strong bonding to the corporation and subordination of individual identity to that of the company. Theory Z organizations have implicit, not explicit, control mechanisms such as peer and group pressure. The norms of the particular corporate culture also provide additional implicit controls. Japanese companies are wellknown for their collective decision-making and responsibility at all levels.

Theory Z management emphasizes a high degree of cross-functionality for all of its workers. Specialization is discouraged. Most top Japanese managers have worked in all aspects of their business from the production floor to sales and marketing. This is also true within functional groups. For example, assemblers will be cross-trained to operate any machine on the assembly floor. Theory Z employers are notoriously slow in giving promotions, and most Japanese CEOs are over age 50.

The purpose of this litany of alphabetic management styles is not to promote one over another; in fact, I don't recommend adopting any of these naively. But many individual team members and managers will exhibit some behaviors from one of the above styles, and it is helpful to know what makes them tick. Finally, certain individuals may prefer to be managed as a Theory X or Theory Y type (Theory Z is less likely in this case), and it is good to be able to recognize the signs. Moreover, some companies might be implicitly based on one style or another.

What is Theory W?

Theory W is a software project management paradigm developed by Boehm [1989], which focuses on the following for each project:

- establishing a set of win-win preconditions
- structuring a win-win software process
- structuring a win-win software product

What does it mean to establish a set of win-win preconditions?

This means recognizing that the best working relationships are those in which everyone "wins." Zero-sum, win-lose, or lose-win situations can leave one or both parties bitter.

Win-win solutions can be sought as follows. First, recognize that everyone wants to win. Then, understand what constitutes a winning situation for each individual. Money, power, and recognition contribute to winning conditions for most people, but there are other, more subtle, conditions such as job satisfaction, a feeling of belonging, and moral fulfillment.

Next, establish reasonable expectations. The importance of setting reasonable and mutually fulfilling expectations in every aspect of human relations can't be overemphasized. Then, ensure that task assignments match the win conditions.

Finally, provide an environment that supports the fulfillment of the win conditions. This can take a variety of forms but might include such things as financial incentives, group activities, and communication sessions to head off problems.

What does it mean to structure a win-win software process?

It means setting up a software process that will lead to success. This includes establishing a realistic process plan based on some standard methodology. This methodology may be internal and company-wide, or off-the-shelf.

It is also important to use the project/management plan to control the project. Too often, managers develop a project plan to sell the job to senior management or the customer, and then throw the plan away once it is approved. Therefore, be sure to use and maintain the project plan throughout the life of the project.

Project managers also need to monitor the risks that have been described as they can lead to win-lose or lose-lose situations. Thus, risks should be identified and eliminated at the earliest opportunity.

Keeping people involved is essential. It helps team members feel a part of the project and improves communications. Besides, listening to team members can reveal great ideas.

What does structuring a win-win software product mean?

This refers to the process of specification writing by matching the users' and maintainers' win conditions. This process also requires careful and honest expectation setting.

What is Principle Centered Leadership?

All of the management approaches discussed thus far focus on organizational frameworks for management. Principle Centered Leadership focuses on the behavior of the manager as an agent for change [Covey 1991]. Some management theorists hold that motivating team members by example and leadership, and not through hierarchical application of authority, is much more effective (manage things, but lead people). A key concept in Principle

Centered Leadership is that the best managers are leaders and that the only way to affect change is by the managers changing themselves first.

Principle Centered Leadership recognizes that principles are more important than values. Values are society-based and can change over time and differ from culture to culture. Principles are more universal, more lasting. Think of some of the old principles like the "Golden Rule." That is, treat others as you would like to be treated. These kinds of principles are timeless and transcend cultures.

In fact, there is a great deal of similarity in Principle Centered Leadership and Theory W, with Principle Centered Leadership being much more generic.

What is management by sight?

Also known as management by walking around, this is not really a full-bodied management approach, but rather a sub-strategy for the approaches already discussed. This approach is people-oriented because it requires the manager to be very visible and to interact with staff. Interacting with staff at all levels is a good way for managers to collect important information about the project and people in their care.

Management by sight is obvious. The manager uses observation and visibility to provide leadership, to monitor the situation, and even to control when necessary. In general, it is advisable to incorporate this strategy into any management approach.

What is management by objectives?

Management by objectives (MBO) is another sub-strategy that can be used in conjunction with any other management approach. MBO involves managers and subordinates jointly setting carefully structured objectives with measurable outcomes and rewards.

Coupled with periodic reviews to measure progress, MBO has the effect of positive reinforcement of desired performance.

For example, a manager may agree with the team member responsible for writing a section of the SDD to complete the task by a certain date (provisos can be made for various inevitable distractions that will appear). In return, time off might be granted for meeting the goal; more time off for early completion. The scenario becomes somewhat more complex when the other 10 things for which the team member is responsible are factored in (for example, producing other reports, attending meetings, working on another project simultaneously). Process tracking tools are very helpful in this case. But the real keys to MBO are setting reasonably aggressive goals and having a clear means of measuring success.

7.2.2 Dealing with Problems

How do I deal with difficult people?

Whether they are subordinates, peers, or superiors, dealing with difficult people is always a challenge. The first thing to do is to avoid forming an opinion too soon. Never attribute some behavior to malice when a misunderstanding could be the reason. Almost without exception, taking the time to investigate an issue and to think about it calmly is superior to reacting spontaneously or emotionally.

Whatever management style is employed, the manager should make sure that the focus is on issues and not people. Managers should avoid the use of accusatory language such as telling someone that he is incompetent. The manager should focus, instead, on his feelings about the situation.

Make sure that all sides of the story are listened to when arbitrating a dispute before forming a plan of resolution. It is often said that there are three sides to an issue, the sides of the two opponents and the truth, which is somewhere inbetween. While this is a cliché, there is much truth to it.

The manager should always work to set or clarify expectations. Management failures, parental failures, marital failures, and the like are generally caused by a lack of clear expectations. The manager should set expectations early in the process, making sure that everyone understands them. He should continue to monitor the expectations and refine them if necessary.

Good team chemistry can be fostered through mentoring and most of the best managers fit the description of a mentor. The behaviors already described are generally those of someone who has a mentoring personality.

Finally, the manager should be an optimist. No one chooses to fail The manager should always give people the benefit of the doubt and work with them.

Is that it? Can't you give me a playbook for handling difficult situations?

Team management is a complex issue and there are many books on the subject and a great deal of variation in how to deal with challenging situations. Table 7.2 is a summary of various sources of conflict and a list of things to try to deal with those conflicts.

7.2.3 Hiring Software Engineering Personnel

We need to hire more software people. How should I approach this task?

First, really consider whether you need another person on the team. Remember Brooks' admonition that adding more people to an already late project will make it later. If you need to hire more staff, say, because the scope of work has expanded or you have more work than the current staff can handle, remember Boehm's [1989] five staffing principles:

The principle of top talent: Use better and fewer people.

The principle of job matching: Fit the tasks to the skills and motivation of the people available (remember this when we talk about outsourcing/offshoring).

TABLE 7.2

Various Sources of Conflict and Suggestions on How to Manage Them Based on Analysis of Experience Reports from 56 Companies

	Sources of Conflict	Managing Conflict
Processes	Scarce resource of time	Employ time management Plan for schedule overruns Manage effect of schedule changes Learn from project experience
	User vs. technical requirements	Identify common goals Align individual goals with process metrics Value team more than individual success
People	Disagreement	Apply team building principles Train in conflict resolution Sponsor group activities Support informal social contact
	Personalization of code	Understanding of one another's point of view
Organization	Power and politics	Structure for success Co-locate teams Integrate development/testing functions Instill ownership
	Management behavior	Get leadership involved Create a collaborative atmosphere Model effective conflict management

Source: Adapted from [Stelzer 1998].

The principle of career progression: An organization does best in the long run by helping its people to self-actualize.

The principle of team balance: Select people who will complement and harmonize with one another.

The principle of phaseout: Keeping a misfit on a team doesn't benefit anyone.

I want to select the right people, but how is it done in the software industry?

Companies use a variety of testing instruments to evaluate job applicants for skills, knowledge, and even personality. Candidate exams range widely from programming tests, general knowledge tests, situation analysis, and "trivia" tests that purport to test critical thinking skills. Many times domain-specific tests are given. Some companies don't test at all.

In some companies, the test is written, administered, and graded by the Human Resources department. Elsewhere, the technical line managers administer the test. In still other places, the tests are organized and delivered by "peer" staff — those who would potentially work with the candidate and who seem to be in the best position to determine which technical skills are relevant. Finally, some companies outsource this kind of testing.

During the dot.com era, a number of Web-based companies emerged which provided online testing. These companies wrote and managed a test databank and provided incentives for candidates to visit a Web site and take the test. For a fee, subscribing companies could access the database and interview those candidates who had achieved a certain score or answered certain questions correctly on the test.

Do these tests really measure the potential success of the software engineer?

There seems to be no consensus, even among those who study human performance testing, that these tests strongly correlate with new employee success.

One thing a company could do to test the efficacy of a knowledge/skills exam is to give it to a group of current employees who are known to be successful, and to another group who are less successful. Correlating the results could lead to a set of questions that "good" employees are more likely to get right (or wrong). Alternatively, a company could survey those employees who were fired for their poor skills to identify any common trends. Clearly though, while such studies might be interesting, they would be nearly impossible to conduct and, in any case, it is unclear if these tests measure what a hiring manager really wants to know.

Perhaps skill or knowledge testing are not what is needed. In many cases, a failure in attitude leads to performance shortfalls. Perhaps then some sort of assessment of potential to get along with others might be needed. Some organizations will measure a candidate's compatibility with the rest of the team by testing "emotional intelligence," or personality. The idea is to establish the "style" of existing team members based on their personalities, then look to add someone whose style is compatible.

In any case, it is questionable as to whether these kinds of tests really do lead to better hires. Unfortunately, it is also hard to gauge a person's fit with the team based on a series of interviews with team members and the obligatory group lunch interview. Anyone can role-play for a few hours in order to get hired, and most interviewers are not that well-trained to see through an act.

You don't seem to like these tests. How do I assess the potential of a candidate besides checking references?

If you must use an assessment of "intelligence," college grades or graduate record examination scores may be used. If the real goal is determine programming prowess, checking grades in programming courses might be helpful. Better still, the manager can ask the candidate to bring in a sample of some code she developed and discuss it. If the code sample is too trivial or the candidate struggles to explain it, it is likely she didn't write it or understand it that well.

Other skills can be tested in this manner as well. For example, if the job entails writing software specifications, design, or test plans, ask the candidate to provide a writing sample. It is possible the candidate does not have samples from his current or past job for proprietary or security reasons, but he should still be able to talk about what he did without notes. If he can recount the project in detail, then he probably knows what he is talking about.

How do you measure the candidate's compatibility with existing team members?

Many companies use personality tests, and in some cases this might work. But there is no magic test here. The manager has to do her homework. She must spend time with the candidate in a variety of settings including having lunch (you can learn a lot about a person from his manners, for example). Make sure that whomever will be working closely with this person is on the hiring/interview team and gets to meet the candidate. Also make sure that everyone has been trained on what to look for during an interview. Then get together and compare notes right after the candidate leaves.

How should I reference-check a potential hire?

Most people don't do a good job of checking references. Here are some simple guidelines for reference checking:

Ask legal questions. There are many questions that cannot be asked and a human resources or legal advisor should be consulted before writing the interview questions.

Evaluate the references the candidate has given you. If they have difficulty providing references, if none is a direct supervisor (or subordinate), then this may indicate a problem. If the reference barely knows the candidate or simply worked in the same building, then regardless of his opinions, they cannot be weighed heavily.

Be sure you check at least three references. It is harder to hide any problems this way. Be sure to talk to supervisors, peers, and subordinates. A team member has to be able to lead and be led.

Ask a "hidden" reference. I always try to track down someone who would know the candidate, but who is not on his reference list. To avoid potential problems with his current employer, I go back to one of his previous employers and track down someone who knew him well there. This strategy is very effective to find out if the candidate has any hidden issues.

Take good notes and ask follow-up questions. Many references are reluctant to say bad things about people even if they do not believe the person is a strong candidate. By listening carefully to what references do and do not say, the real message will come through.

Be sure to ask a broad range of questions, and questions that encourage elaboration (as opposed to "yes" or "no" questions). For example, some of the following might be helpful:

Describe the candidate's technical skills.

Describe a difficult situation the candidate encountered and how he dealt with it.

Describe the candidate's interpersonal skills.

Why do you think the candidate is leaving the company?

Describe the kind of work environment in which the candidate would thrive.

Describe the current work environment in which the candidate works.

Describe the contributions of the candidate in the capacities with which you are familiar.

Describe the kind of manager you think would be best for the candidate.

In summary, it is crucial that the hiring manager and hiring/interview team learn the art of interviewing and background checking. This is more likely to lead to the right fit than a series of tests. Relying on "trivia tests" is risky at best and you might lose a good employee in the process.

7.2.4 Agile Development Teams

How do I manage agile development teams?

To answer this question, I will tell you about the Agile Manifesto. The Agile Manifesto is a document that lays the groundwork for all agile development methodologies. Its authors include many notable pioneers of object technology including: Kent Beck (JUnit with Eric Gamma), Alistair Cockburn (Crystal), Ward Cunningham (Wiki, CRC cards), Martin Fowler (many books on XP, patterns, UML), Robert C. Martin (Agile, UML, patterns), and Ken Schwaber (Scrum).

The Agile Manifesto

- Our highest priority is to satisfy the customer through early and continuous delivery of valuable software.
- *Welcome changing requirements, even late in development.* Agile processes harness change for the customer's competitive advantage.
- Deliver working software frequently, from a couple of weeks to a couple of months, with a preference to the shorter timescale.
- Business people and developers must work together daily throughout the project.
- Build projects around motivated individuals. Give them the environment and support they need, and trust them to get the job done.
- The most efficient and effective method of conveying information to and within a development team is face-to-face conversation.
- Working software is the primary measure of progress.
- Agile processes promote sustainable development. *The sponsors, developers, and users should be able to maintain a constant pace indefinitely.*
- Continuous attention to technical excellence and good design enhances agility.

- Simplicity — the art of maximizing the amount of work not done — is essential.
- The best architectures, requirements, and designs emerge from self-organizing teams.
- *At regular intervals, the team reflects on how to become more effective,* then tunes and adjusts its behavior accordingly [Beck et al. 2006].

Notice the portions in italics (my emphasis), which provide specific management advice.

OK, so what does the Agile Manifesto have to do with managing agile teams?

The Agile Manifesto is ready-made advice for managers. It implies that managing agile teams is fun (Theory Y), and that the best outcomes arise from giving project teams what they need and leaving them alone.

Does this approach always work?

No. Agile methods require much more autonomy than many managers are willing to give. More importantly, however, not everyone fits the agile methodology — a Theory X type worker will not thrive well with it. Finally, as we discussed in Chapter 2, agile methodologies do not work in every environment or with every project.

Do you have some more specific advice for managing agile teams?

Don Reifer [2002b] gives some excellent advice. He recommends that you clearly define what "agile methods" means upfront because, as we saw, there are many misconceptions. Then build a business case for agile methods using "hard" data. Reifer also notes that when adopting agile methods, recognize that you are changing the way your organization does business.

So, in order to be successful you need to provide staff with support for making the transition. That support should include startup guidelines, "how to" checklists, and measurement wizards; a knowledge base of past experience accessible by all; and education and training, including distance education and self-study courses [Reifer 2002b].

7.3 Project Management Basics

What is a project?

A project is a set of tasks with a defined beginning and end. Without a defined beginning, there is no way to begin measuring progress. Without a defined end, there is no way to determine if the project has been completed, and thus progress towards completion cannot be measured. The simple project definition is recursive in that any project probably consists of more than one sub-project.

What makes a software project different from any other kind of project?

Throughout the text, various properties of software have been discussed. What has been infrequently noted, however, is that the things that make software different from other types of endeavors also make it harder to manage the software process. For example, unlike hardware to a large extent, software designers build software knowing that it will have to change. Hence, the designer has to think about both the design and redesign. That adds a level of complication.

Of course, software development involves novelty, which introduces uncertainty. It can be argued that there is a higher degree of novelty in software than in other forms of engineering.

The uniqueness of software project management is intensified by a number of specialized activities. These include:

- the process of software development
- the complex software maintenance process
- the unique, and not-well-evolved, process of verification and validation
- the interplay of hardware and software
- the uniqueness of the software

Is software project management similar to systems project management?

Many software engineering project management activities are different from those needed for software project management. These are summarized in Table 7.3. This framework provides a model of discussion for the rest of this chapter.

What does the software project manager control?

Software project managers may have one or more of the following elements under their control:

- resources
- schedule
- functionality

Note that I say "may have" control over. It could be that the project manager controls none or only one of these. Obviously, the extent to which the manager has control indicates the relative freedom to maneuver.

Note that there is one aspect here that every manager can control — himself. That is, you have control over your reactions to situations and your attitude. These must be positive if you expect positive responses from others.

TABLE 7.3

Software Process Planning vs. Project Planning

Software Engineering Planning Activities	Software Project Management Planning Activities
Determine tasks to be done	Determine skills needed for the task
Establish task precedence	Establish project schedule
Determine level of effort in person months	Determine cost of effort
Determine technical approach to solving problem	Determine managerial approach to monitoring project status
Select analysis and design tools	Select planning tools
Determine technical risks	Determine management risks
Determine process model	Determine process model
Update plans when the requirements or development environment change	Update plans when the managerial conditions and environment change

Source: Adapted from Thayer, R.H., Software system engineering: a tutorial, *Computer*, 35(4), 68–73, 2002.

What do you mean by resources?

Resources can include software, equipment, and staffing, and money to acquire more of them. There are always financial limitations, and generally these are fixed prior to the start of the project. Many times the financing constraints change during the project.

What about the schedule?

The manager should have some control over the schedule. Even if the delivery date of the product is hard, there should be some flexibility in the schedule that does not change the delivery date.

What about product functionality?

The product functionality may or may not be controllable. Often when negotiating a project, the project manager cannot increase costs or reduce delivery time, but he can decrease product functionality in order to meet a customer's budget or schedule.

How does the project manager put all of these control factors together?

Generally, in terms of controlling the project, the manager must understand the project goals and objectives. Next, the manager needs to understand the constraints imposed on the resources. These include cost and time limitations, performance constraints, and available staff resources. Finally, the manager develops a plan that enables her to meet the objectives within the given constraints.

Of course, monitoring and control mechanisms must be in place including metrics. The manager should be prepared to modify the plan as it progresses. These modifications need to be made to the plan, and then team members can make adjustments as necessary and appropriate. Finally, a calm, productive, and positive environment is desirable to maximize the performance of the team and to keep the customer happy and confident that the job is being done right.

7.4 Tracking and Reporting Progress

What is a work breakdown structure and why is it important to project tracking?

The work breakdown structure (WBS) is used to decompose the functionality of the software in a hierarchical fashion. The WBS can be used for costing and project management and it forms the basis for process tracking and cost determination. The WBS consists of an outline listing of project deliverables (or phases of the project) organized hierarchically.

Figure 7.1 illustrates a simple example for the software engineering effort for the baggage inspection system. A portion of the SRS is shown.

Each organization uses its own terminology for classifying WBS components according to their level in the corporate hierarchy. The WBS may also be organized around deliverables or phases of the project life cycle. In this case, higher levels generally are performed by groups while the lowest levels are performed by individuals. A WBS that emphasizes deliverables does not necessarily specify activities.

1.1 Software Systems Engineering
- 1.1.1 Support to Systems Engineering
- 1.1.2 Software Engineering Trade Studies
- 1.1.3 Requirement Analysis (System)
- 1.1.4 Requirement Analysis (Software)
- 1.1.5 Equations Analysis
- 1.1.6 Interface Analysis
- 1.1.7 Support to System Test

1.2 Software Development
- 1.2.1 Deliverable Software
  - 1.2.1.1 Requirement Analysis
  - 1.2.1.2 Architectural Design

FIGURE 7.1
High-level work breakdown structure for the baggage inspection system.

What is the level of detail of the tasks in the WBS?

Breaking down a project into its component parts facilitates resource allocation and the assignment of individual responsibilities. But care should be taken to use a proper level of detail when creating the WBS. A very high level of detail is likely to result in micromanagement. Too low a level of detail and the tasks may become too large to manage effectively. Generally, I like to define tasks so that their duration is between several days and a few months.

What is the WBS's role in project planning?

The work breakdown structure is the foundation of project planning. It is developed before dependencies are identified and activity durations are estimated. The WBS can also be used to identify the tasks to be used in other project management tools.

Are there any drawbacks to the traditional WBS?

Yes, the WBS is closely associated with the waterfall model, although it can be used with other life-cycle models. The WBS can have the tendency to drive the software architecture; for instance, the modular decomposition looks exactly like the WBS.

Are there any alternatives to using the WBS?

No. Some project managers try to go directly to scheduling without using a WBS. In other cases, project managers try to utilize use cases as the basis of project management, but I do not recommend this approach.

How is work and progress tracked in software projects?

Tracking progress is important for identifying problems early, for reporting purposes, and to perform appropriate resource allocation and reallocation as required. Three tools that can help the software project manager to measure progress of a project are:

the Gantt chart
the critical path method (CPM)
the program evaluation and review technique (PERT)

There are numerous commercial implementations of these tools, which typically can convert from one to the other and integrate with many popular word processing, spreadsheet, and presentation software.

What is a Gantt chart?

Henry Gantt developed the Gantt chart during World War I for use as a planning tool. This widely used tool is simple in that it lists project tasks in a sequential and parallel fashion.

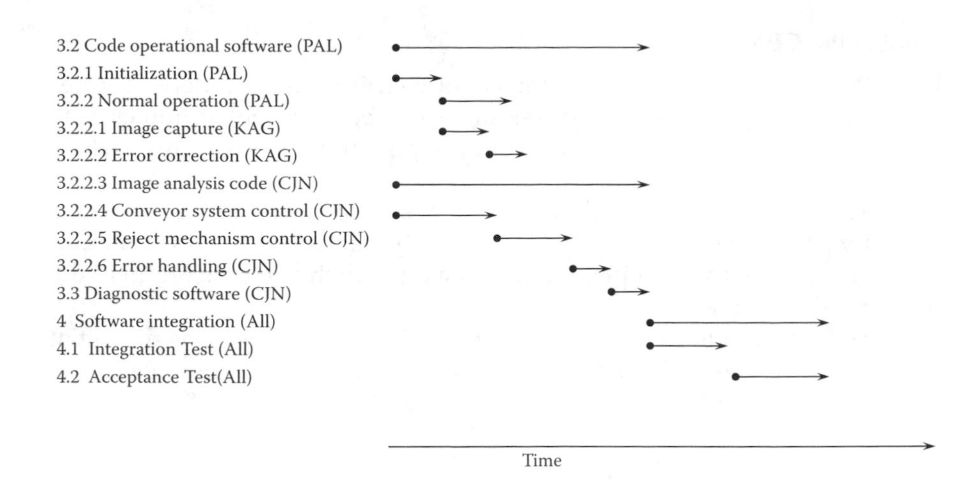

FIGURE 7.2
Partial Gantt chart for the baggage inspection system.

What does the Gantt chart look like?

Consider the Gantt chart shown in Figure 7.2 for the baggage inspection system. Project tasks are listed along the left-hand side of the chart in a hierarchical fashion. If a work breakdown structure was used in the SDD, then it can be transferred to the chart.

A timeline is drawn along the bottom edge of the chart. Here the time units are omitted, but would usually be represented by tick marks in units of days, weeks, or months. Each project subtask activity is represented by a directed arrow. The starting point of the arc is placed at the point in the timeline where the task would commence. Project durations are represented by the length of the arcs. Personnel are listed next to the project activity on the left-hand side. Milestones can be marked and task slippage can be denoted by dashed lines in the activity arcs. The chart is updated as the project unfolds.

It can be seen from Figure 7.2 that parallel tasks can be identified and sequencing can be easily depicted. Task assignments can be made by writing the name of the responsible person next to each task. PAL, CJN, and KAG are the initials of the persons assigned to the tasks, "All" represents that all team members are involved in the task.

Can the Gantt chart be used for large projects?

For larger projects, the tasks can be broken down into subtasks having their own Gantt charts to maintain readability.

How can the Gantt chart be used for ongoing project management?

The strength of the Gantt chart is its capability to display the status of each activity at a glance. So long as the chart is a realistic reflection of the situation, the manager can use it to track progress, adjust the schedule, and perhaps most importantly, communicate the status of the project.

What is the CPM?

The CPM is an improvement on the Gantt chart in that task dependencies can be more easily depicted and task times can be represented numerically rather than visually. The method was developed in the 1950s by researchers at DuPont and Remington Rand.

The CPM chart is essentially a precedence graph connecting tasks and illustrating their dependencies along with the budgeted completion time and maximum cumulative completion time along the path from the origin to the current task (see Figure 7.3).

For example, in Figure 7.3 the tasks are A, B, C, …, K. Task A is the initial task followed by tasks B and C, which cannot start until A is completed. The time to complete task A is 1 week and for B and C 2 weeks.

What are the steps in CPM planning?

First, specify the individual activities, which can be obtained from the work breakdown structure. This listing can be used as the basis for adding sequence and duration information in later steps.

Next, determine the sequence of those activities, including any dependencies. Note that some activities are dependent upon the completion of others. A listing of the immediate predecessors of each activity is useful for constructing the CPM network diagram.

Now draw a network diagram. CPM was originally developed as an activity on node network, but some project planners prefer to specify the activities on the arcs.

Next, estimate the completion time for each activity. The time required to complete each activity can be estimated using past experience or the estimates of knowledgeable persons. CPM is a deterministic model that does not take into account variation in the completion time, so only one number is used for an activity's time estimate.

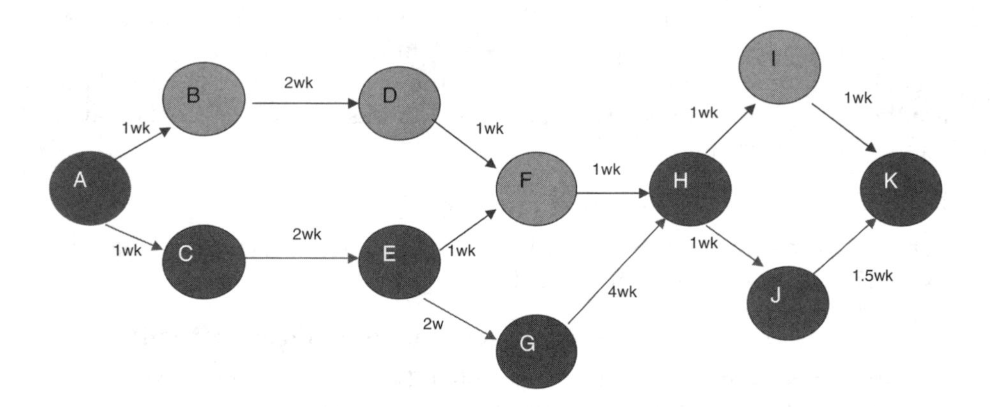

FIGURE 7.3
A generic CPM chart with the critical path highlighted.

Then identify the critical path, which is the path through the project network that has the greatest aggregate completion time.

Finally, update the CPM diagram as the project progresses because the actual task completion times will then be determined and the network diagram can be updated to include this information. A new critical path may emerge, and structural changes may be made in the network if project requirements change.

Can you illustrate the technique using the baggage inspection system?

Returning to the baggage inspection system example, consider the tasks in the work breakdown structure by their numerical coding shown in the Gantt chart. These tasks are depicted in Figure 7.4.

Here tasks 3.2.1, 3.2.2.3, and 3.2.2.4 can begin simultaneously. Assume that task 3.2.1 is expected to take four time units (days). Notice that, for example, the arc from task 3.2.1 is labeled with "4/4" because the estimated time for that task is four days, and the cumulative time along that path up to that node is four days. Looking at task 3.2.2.1, which succeeds task 3.2.1, we see that the edge is labeled with "4/8". This is because a completion time for task 3.2.2.1 is estimated at four days, but the cumulative time for that path (from tasks 3.2.1 through 3.2.2.1) is estimated to be a total of eight days.

Moving along the same path, task 3.2.2.1 is also expected to take four days, so the cumulative time along the path is eight days. Finally, task 3.2.2.2 is expected to take three days, and hence the cumulative completion time is

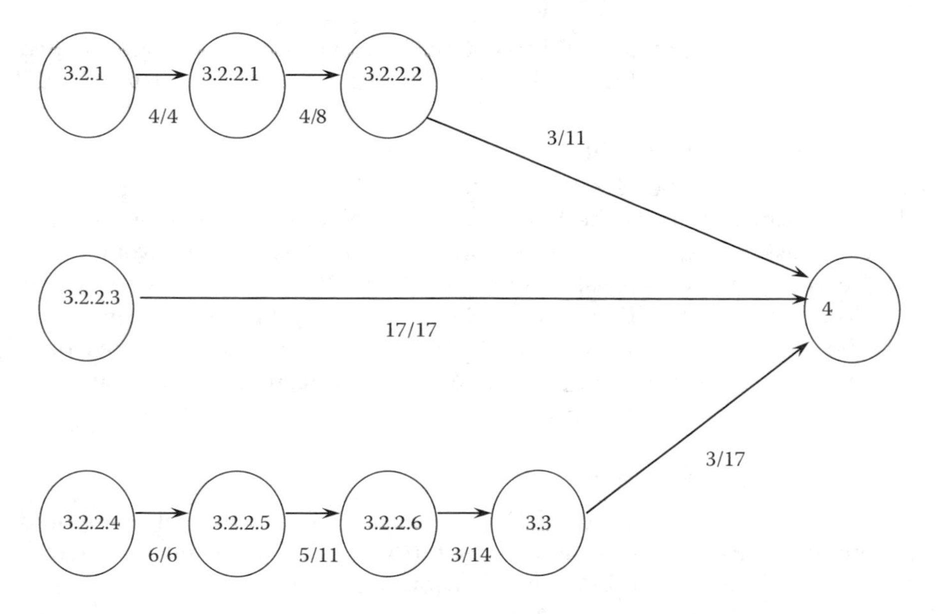

FIGURE 7.4
Partial CPM corresponding to the baggage inspection system Gantt chart shown in Figure 7.2.

11 days. On the other hand, task 3.2.2.3 is expected to take 17 days. The task durations are based on either estimation or using a tool such as COCOMO, which will be discussed later. If the Gantt chart accompanies the CPM diagram, the task durations represented by the length of the arrows on the Gantt chart should correspond to those labeled on the CPM chart.

Moving along the last path at the bottom of Figure 7.4, it can be seen that the cumulative completion time is 17. Therefore, in this case, the two lower task paths represent critical paths. Hence, only by reducing the completion time of both lower task paths can the project completion be accelerated.

Are there downsides to using CPM?

CPM was developed for complex but fairly routine projects with minimal uncertainty in the project completion times. For less routine projects, there is more uncertainty in the completion times and this uncertainty limits the usefulness of the deterministic CPM model. An alternative to CPM is the PERT project planning model, which allows a range of durations to be specified for each activity.

What is PERT?

PERT was developed by the Navy and Lockheed (now Lockheed Martin) in the 1950s, around the same time as CPM. PERT is identical to CPM topologically, except that PERT depicts optimistic, likely, and pessimistic completion times along each arc.

How do you build the PERT diagram?

The steps are the same as for CPM except that when you determine the estimated time for each activity, optimistic, likely, and pessimistic times are determined.

Can you show me an example?

In Figure 7.5, it can be seen that the topology is the same as that for CPM. Here the triples indicate the best, likely, and worst-case completion times for each task. These times are estimated, as in CPM, either through best engineering judgment or using a tool like COCOMO. Adding these triples vectorially yields the PERT chart in Figure 7.6. The aggregated times can now be seen along the arcs, providing cumulative best, likely, and worst-case scenarios. This provides even more control information than the Gantt or CPM project representations.

Are there any downsides to using PERT?

Yes. For example, the activity time estimates are somewhat subjective and depend on judgment (that is, guessing). Further, PERT assumes that the task completion time is given by a beta distribution of the form

$$f(t) = kta - 1(1 - t)b - 1$$

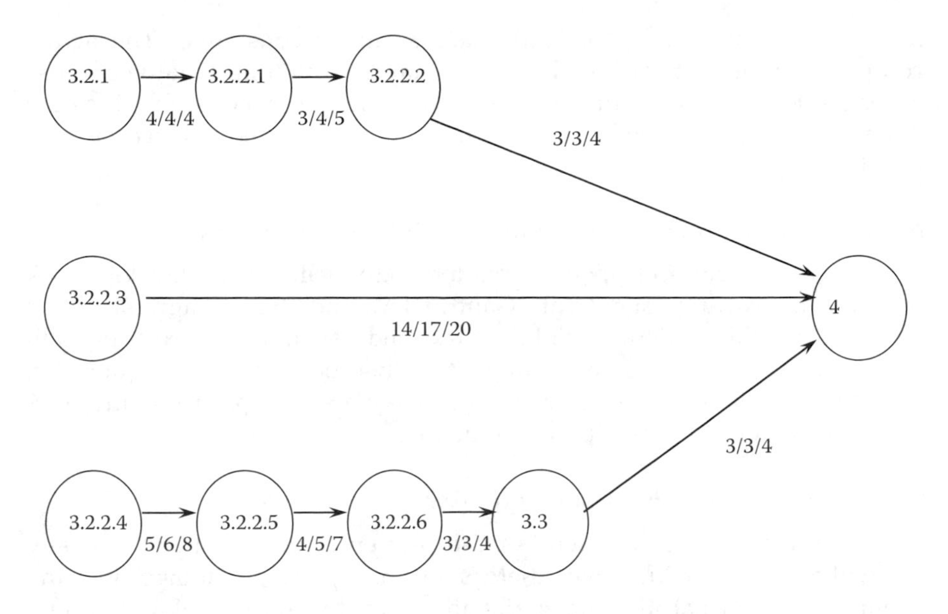

FIGURE 7.5
Partial PERT chart for the baggage inspection system showing best/likely/worst case completion times for each task.

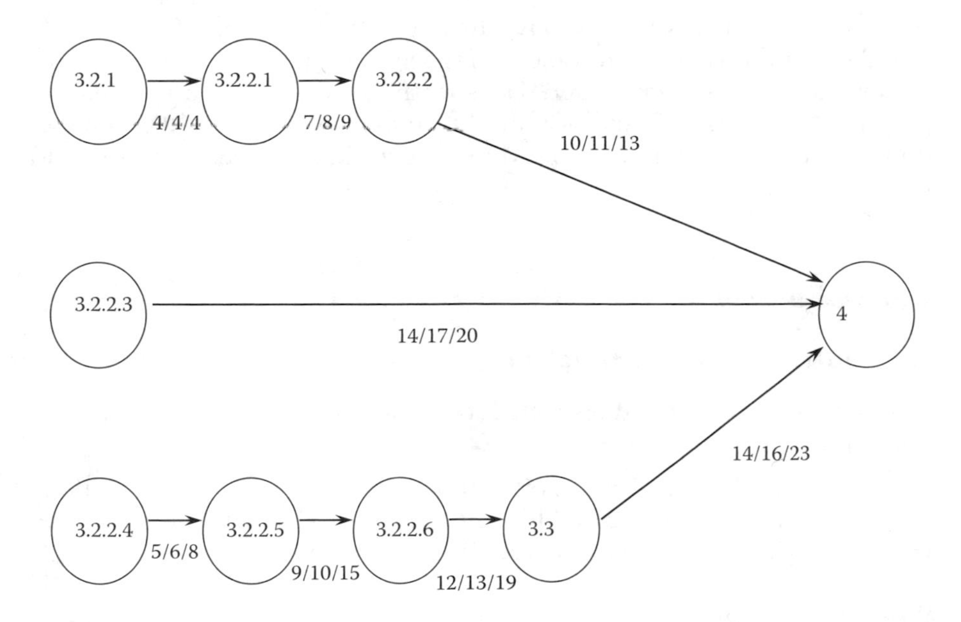

FIGURE 7.6
Partial PERT chart for the baggage handling system showing cumulative best/likely/worst case completion times for each task.

for the time estimates (k, a, and b are arbitrary constants). The actual completion time distribution, however, may be different. Finally, because other paths can become the critical path if their associated activities are delayed, PERT consistently underestimates the expected project completion time.

Are commercial products available for building these charts?

Several commercial and open source tools are available to develop work breakdown structures and create Gantt, CPM, and PERT diagrams. Some tools also provide to-do lists, linking tasks and dependencies, role- or skill-based tracking, and resource leveling. Still other tools provide collaboration features such as project status reports accessible via a Web page and integrated e-mail or threaded discussion boards.

Can you recommend the best tool to use?

It is not my place to recommend one implementation over another — any one will probably do. Moreover, there is no "managerial advantage" in using one tool or another. The "managerial advantage" is in the skill and knowledge of the person using the tool.

Is becoming an expert in using the project planning tools the key to being a good software project manager?

Absolutely not. Frankly it is naïve to imply that mastery of a tool is a foundation for excellence in some profession. Does mastery of a word processor make you a great writer? Does learning how to use a spreadsheet program make you a financial whiz? Of course not. You need the tools to do your job, of course, but you need to have the knowledge and wisdom to use those tools appropriately.

7.5 Software Cost Estimation

Are there well-known and respected tools for doing software project cost estimation?

One of the most widely used software modeling tools is Boehm's [1981] constructive cost model (COCOMO), first introduced in 1981. There are three versions of COCOMO: basic, intermediate, and detailed.

What is basic COCOMO?

Basic COCOMO is based on thousands of lines of deliverable source instructions. In short, for a given piece of software, the time to complete is a function of L, the number of lines of delivered source instructions (KDSI), and two additional parameters, *a* and *b*, which will be explained shortly. This is the effort equation for the basic COCOMO model:

$$T = aL^b \tag{7.1}$$

Dividing T by a known productivity factor, in KDSI per person month, yields the number of person months estimated to complete the project.

The parameters *a* and *b* are a function of the type of software system to be constructed.

For example, if the system is organic (i.e., one that is not deeply embedded in specialized hardware), then the following parameters are used: $a = 3.2$, $b = 1.05$. If the system is semidetached (i.e., partially embedded), then the following parameters are used: $a = 3.0$, $b = 1.12$.

Finally, if the system is embedded like the baggage inspection system, then the following parameters are used: $a = 2.8$, $b = 1.20$. Note that the exponent for the embedded system is the highest, leading to the longest time to complete for an equivalent number of delivered source instructions.

Can you give me an example using COCOMO?

Suppose it is estimated somehow that the baggage inspection system will require 40 KDSI of new C code to complete. Hence, an effort level estimate of:

$$T = 2.8 \cdot (40K)^{1.2} = 234K$$

is obtained using COCOMO.

Suppose, then, it is known that an efficient programmer can generate 2000 LOC per month. Then, superficially at least, it might be estimated that the project would take approximately 117 person months to complete. Not counting dependencies in the task graph, this implies that a five-person team would take approximately 20 months to complete the project. It would be expected, however, that more time would be needed because of task dependencies (identified, for example, using PERT).

Where do the source code estimates come from?

These can come from function, feature, or use case point calculations, be based on an analysis of similar types of recently completed projects, or be provided by expert opinions.

Should I use more than one estimate?

Yes. You should use at least two methods to estimate KDSI. You can take a weighted average of the two with the weights based on your certainty of the estimate.

What about the intermediate and detailed COCOMO models?

The intermediate or detailed COCOMO models dictate the kinds of adjustments used. Consider the intermediate model, for example. Once the effort level for the basic model is computed based on the appropriate parameters and number of source instructions, other adjustments can be made based on additional factors.

In this case, for example, if the source code estimate consists of design-modified code, code-modified and integration-modified rather than straight code, a linear combination of these relative percentages is used to create an adaptation adjustment factor as follows.

Adjustments are then made to *T* based on two sets of factors, the adaptation adjustment factor, *A*, and the effort adjustment factor, *E*.

What is the adaptation adjustment factor?

The adaptation adjustment factor is a measure of the kind and proportion of code that is to be used in the system, namely, design-modified, code-modified, and integration-modified. The adaptation factor, *A*, is given by Equation 7.2.

$$A = 0.4\ (\%\ \text{design-modified}) + .03\ (\%\ \text{code-modified}) + 0.3\ (\%\ \text{integration-modified}) \tag{7.2}$$

For new components $A = 100$. On the other hand, if all of the code is design-modified, then $A = 40$ and so on. Then the new estimation for delivered source instructions, *E*, is given as

$$E = L \cdot A/100 \tag{7.3}$$

What is the effort adjustment factor?

An additional adjustment, the effort adjustment factor, can be made to the number of delivered source instructions based on a variety of factors including:

- product attributes
- computer attributes
- personnel attributes
- project attributes

Each of these attributes is assigned a number based on an assessment that rates them on a relative scale. Then, a simple linear combination of the attribute numbers is formed based on project type. This gives a new adjustment factor, E'.

The second adjustment, effort adjustment factor, E'', is then made based on the formula

$$E'' = E' \cdot E \tag{7.4}$$

Then the delivered source instructions are adjusted, yielding the new effort equation:

$$T = aE''^b \tag{7.5}$$

The detailed model differs from the intermediate model in that different effort multipliers are used for each phase of the software life cycle.

What do these adjustment factors look like?

Table 7.4 lists the adjustment factors corresponding to various product attributes.

How widely used is COCOMO?

COCOMO is widely recognized and respected as a software project management tool. It is useful even if the underlying model is not really understood. COCOMO software is commercially available and can be found on the Web free.

What are the downsides to using COCOMO?

One drawback is that the model does not take into account the leveraging effect of productivity tools. The model also bases its estimation almost entirely on LOC, not on program attributes, which is something that FPs do. FPs, however, can be converted to source code estimates using standard conversion formulas, as was shown in Chapter 6.

What is COCOMO II?

COCOMO II is a major revision of COCOMO that is evolving to deal with some of the previously described shortcomings. For example, the original COCOMO 81 model was defined in terms of delivered source instructions. COCOMO II uses the metric SLOC instead of delivered source instructions. The new model helps better accommodate more expressive modern languages as well as software generation tools that tend to produce more code with essentially the same effort.

In addition, some of the more important factors that contribute to a project's expected duration and cost are included in COCOMO II as new scale drivers. These five scale drivers are used to modify the exponent in the effort equation:

precedentedness (novelty of the project)
development flexibility
architectural/risk resolution
team cohesion
process maturity

TABLE 7.4

Attribute Adjustment Factors for Intermediate COCOMO

	Very Low	Low	Nominal	High		Very High	Extra High
Product attributes							
Required software reliability	0.75	0.88	1.00	1.15		1.40	
Size of application database		0.94	1.00	1.08		1.16	
Complexity of the product	0.70	0.85	1.00	1.15		1.30	1.65
Hardware attributes							
Run-time performance constraints			1.00	1.11		1.30	1.66
Memory constraints			1.00	1.06		1.21	1.56
Volatility of the virtual machine environment		0.87	1.00	1.15		1.30	
Required turnabout time		0.87	1.00	1.07		1.15	
Personnel attributes							
Analyst capability	1.46	1.19	1.00	0.86	0.71		
Software engineer capability	1.29	1.13	1.00	0.91	0.82		
Applications experience	1.42	1.17	1.00	0.86	0.70		
Virtual machine experience	1.21	1.10	1.00	0.90			
Programming language experience	1.14	1.07	1.00	0.95			
Project attributes							
Use of software tools	1.24	1.10	1.00	0.91	0.82		
Application of software engineering methods	1.24	1.10	1.00	0.91	0.83		
Required development schedule	1.23	1.08	1.00	1.04	1.10		

Note: Entries are empty if they are not applicable to the model.

Source: Adapted from Wikipedia.org.

The first two drivers, precedentedness and development flexibility, describe many of the same influences found in the adjustment factors of COCOMO 81.

What is WEBMO?

WEBMO is a derivative of COCOMO II that is geared specifically to project estimation of Web-based projects (it has been reported that COCOMO is not a good predictor in some cases). WEBMO is based on a different set of predictors, namely,

- number of function points
- number of xml, html, and query language links
- number of multimedia files
- number of scripts
- number of Web building blocks

For WEBMO, the effort and duration equations are:

$$\text{Effort} = A\prod_{i=1}^{9} cd_i(size)^{P_1} \tag{7.6}$$

$$\text{Duration} = B(Effort)^{P_2} \tag{7.7}$$

where A and B are constants, P_1 and P_2 depend on the application domain, and cd_i are cost drivers based on:

- product reliability and complexity
- platform difficulty
- personal capabilities
- personal experience
- facilities
- schedule constraints
- degree of planned reuse
- teamwork
- process efficiency

with qualitative ratings ranging from very low to very high and numerical equivalents shown in Table 7.5 [Reifer 2002a].

TABLE 7.5

WEBMO Cost Drivers and Their Values

	Ratings				
	Very Low	**Low**	**Nominal**	**High**	**Very High**
Cost driver					
Product reliability	0.63	0.85	1.0	1.30	1.67
Platform difficulty	0.75	0.87	1.00	1.21	1.41
Personnel capabilities	1.55	1.35	1.00	0.75	0.58
Personnel experience	1.35	1.19	1.00	0.87	0.71
Facilities	1.35	1.13	1.00	0.85	0.68
Schedule constraints	1.35	1.15	1.00	1.05	1.10
Degree of planned reuse	—	—	1.00	1.25	1.48
Teamwork	1.45	1.31	1.00	0.75	0.62
Process efficiency	1.35	1.20	1.00	0.85	0.65

Source: Reifer, D.J., Estimating web development costs: there are differences, http://www.reifer.com/documents/webcosts.pdf, June 2002.

7.6 Project Cost Justification

Is software an investment or an expense?

It depends on whom you ask. Many software project managers see the investment in new tools or upgrade of old ones as an investment. But the CFO might see the purchase as a pure expense. There is an accounting answer to this question, too, but I don't want to get into the technical details of how accountants determine whether a purchase is an expense item or a capital acquisition.

The point is that many software project managers are being asked to justify their activities and purchases of software and equipment. Therefore, it is in the project manager's best interest to know how to make a business case for the activity.

What is software return on investment (ROI) and how is it defined?

Return on investment (ROI) is a rather overloaded term that means different things to different people. To some it means the value of the software activity at the time it is undertaken. To some it is the value of the activity at a later date. To some it is just a catchword for the difference between the cost of software and the savings anticipated from the utility of that software. Finally, to some there is a more complex meaning.

What is an example of a project ROI justification?

Consider the following situation. A project manager has the option of either purchasing a new testing tool for $250,000 or using the same resources to hire and train additional testers. Currently $1 million is budgeted for software testing. It has been projected that the new testing tool would provide

$500,000 in immediate cost savings by automating several aspects of the testing effort. The effort savings would allow fewer testers to be assigned to the project. Should the manager decide to hire new testers, they would have to be hired and trained (these costs are included in the $250,000 outlay) before they can contribute to the project.* At the end of two years, it is expected that the new testers will be responsible for $750,000 in rework cost savings by finding defects prior to release that would not otherwise be found. The value justification question is "should the project be undertaken or not?" We can answer this question after discussing net present value.

Yes, but how do you measure ROI?

One traditional measure of ROI for any activity, whether software related or not, is given as

$$\text{ROI} = \text{Average Net Benefits}/\text{Initial Costs}$$

The problem with this model for ROI is the accurate representation of average net benefits and initial costs.

OK, so how can you represent net benefit and initial cost?

Commonly used models for valuation of some activity or investment include net present value (NPV), internal rate of return (IRR), profitability index (PI), and payback. We will look at each of these shortly.

Other methods include Six Sigma and proprietary balanced scorecard models. These kinds of approaches seek to recognize that financial measures are not necessarily the most important component of performance. Further considerations for valuing software solutions might include customer satisfaction, employee satisfaction, and so on, which are not usually modeled with traditional financial valuation instruments.

There are other, more complex, accounting-oriented methods for valuing software. Discussion of these techniques is beyond the scope of this text. The references at the end of the chapter can be consulted for additional information; see, for example, [Raffo et al. 1999] and [Morgan 2005].

What is NPV and how can I use it?

NPV is a commonly used approach to determine the cost of software projects or activities. Here is how to compute NPV. Suppose that FV is some future anticipated payoff either in cash or anticipated savings. Suppose r is the discount rate** and Y is the number of years that the cash or savings is expected to be realized. Then the NPV of that payoff is:

$$\text{NPV} = FV/(1 + r)^Y$$

---

* Such a cost is called a "sunken cost" because the money is gone whether one decides to proceed with the project or not.

** The interest rate charged by the U.S. Federal Reserve. The cost of borrowing any capital will be higher than this base rate.

NPV is an indirect measure because you are required to specify the market opportunity cost (discount rate) of the capital involved.

Can you give an example of an NPV calculation for a software situation?

To see how you can use this notion as a project manager, suppose that you expect a programming staff training initiative to cost your company $60,000. You believe that benefits of this improvement initiative are expected to total $100,000 of reduced code rework two years in the future. If the discount rate is 3%, should the initiative be undertaken?

To answer this question, we calculate the NPV of the strategy, taking into account its cost:

$$\text{NPV} = 100{,}000/1.03^2 - 60{,}000 = 34{,}259$$

Because the NPV is positive, the project should be undertaken.

For a sequence of cash flows, CF_n, where $n = 0, \ldots, k$ represents the number of years from initial investment, the NPV of that sequence is

$$NPV = \sum_{n=0}^{k} \frac{CF_n}{(1+r)^n}$$

CF_n could represent, for example, a sequence of related expenditures over a period of time, such as the ongoing maintenance costs or support fees for some software package.

What is the answer to the question of acquiring the testing tool?

To figure out if we need to undertake this project, we assume an annual discount rate of 10% for ease of calculation. Now we calculate the NPV of both alternatives. The testing tool is worth $500,000 today, so its NPV is

$$\text{PV}_{\text{tool}} = \$500{,}000/(1.10)^0 = \$500{,}000$$

To hire testers is worth $750,000 in two years, so its NPV is

$$\text{PV}_{\text{hire}} = \$750{,}000/(1.10)^2 = \$619{,}835$$

Therefore, under these assumptions, the personnel hire option would be the preferred course of action. However, as the projected return goes farther into the future, it becomes more difficult to forecast the amount of the return. All sorts of things could happen — the project could be cancelled, new technology could be discovered, the original estimate of rework could change. Thus, the risk of the project may differ accordingly.

What is an IRR?

IRR is defined as the discount rate in the NPV equation that causes the calculated NPV to be zero. NPV is not the ROI. But the IRR is useful for computing the "return" because it does not require knowledge of the cost of capital.

To decide if we should undertake an initiative, we compare the computed IRR to the return of another investment alternative. If the IRR is very low, then we might simply want to take this money and find an equivalent investment with lower risk (for example, to undertake a different corporate initiative or even to simply buy bonds). But if the IRR is sufficiently high, then the decision might be worth whatever risk is involved.

Can you give an example of an ROI calculation?

Suppose the programming staff training initiative previously discussed is expected to cost $50,000. The returns of this improvement are expected to be $100,000 of reduced rework two years in the future. We would like to know the IRR on this activity.

Here, NPV = $100{,}000/(1 + r)^2 - 50{,}000$. We now wish to find the r that makes the NPV = 0; that is, the "break even" value. Using our IRR equation,

$$r = [100{,}000/50{,}000)]^{1/2} - 1$$

This means $r = 0.414 = 41.4\%$. This rate of return is very high, and we would likely choose to undertake this programming staff training initiative.

What is a PI?

The PI is the NPV divided by the cost of the investment, *I*:

$$\text{PI} = \text{NPV}/I.$$

PI is a "bang-for-the-buck" measure and it is appealing to managers who must decide between many competing investments with positive NPV financial constraints. The idea is to take the investment options with the highest PI first until the investment budget runs out. This approach is not bad but can suboptimize the investment portfolio.

How can using PI suboptimize the decision?

Consider the set of software investment decisions shown in Table 7.6. Suppose the capital budget is $500,000. The PI ranking technique will pick A and B first, leaving inadequate resources for C. Therefore, D will be chosen

TABLE 7.6

A Portfolio of Software Project Investment Decisions

Project	Investment (in hundreds of thousands of dollars)	NPV (in hundreds of thousands of dollars)	PI
A	200	260	1.3
B	100	130	1.3
C	300	360	1.20
D	200	220	1.1

leaving the overall NPV at $610,000. However, using an integer programming approach will recommend taking projects A and C for a total NPV of $660,000.

Should I use PI at all?

Yes, PI is useful in conjunction with NPV to help optimize the allocation of investment dollars across a portfolio of projects.

What is payback?

A funny answer is warranted here, but I will pass on the opportunity. To the project manager, payback is the time it takes to get the initial investment back out of the project. Projects with short paybacks are preferred, although the term "short" is completely arbitrary. The intuitive appeal is reasonably clear: the payback period is easy to calculate, communicate, and understand.

How can payback be applied in a software project setting?

Suppose changing vendors for a particular application software package is expected to have a switching cost of $100,000 and result in a maintenance cost savings of $50,000 per year. Then the payback period for the decision to switch vendors would be two years.

This seems simplistic. Is payback really used?

Yes. Because of its simplicity, payback is the least likely ROI calculation to confuse managers. However, if the payback period is the only criterion used, then there is no recognition of any cash flow, small or large, to arrive after the cutoff period. Furthermore, there is no recognition of the opportunity cost of tying up funds. Because discussions of payback tend to coincide with discussions of risk, a short payback period usually means a lower risk. However, all criteria used in the determination of payback are arbitrary. From an accounting and practical standpoint, the discounted payback is the metric that is preferred.

What is discounted payback?

The discounted payback is the payback period determined on discounted cash flows rather than undiscounted cash flows. This method takes into account the time (and risk) value of money invested. Effectively, it answers the questions "How long does it take to recover the investment?" and "what is the minimum required return?"

If the discounted payback period is finite in length, it means that the investment plus its capital costs are recovered eventually, which means that the NPV is at least as great as zero. Consequently, a criterion that says to go ahead with the project if it has *any* finite discounted payback period is consistent with the NPV rule.

Can you give an example?

In the previous PI example, there is a switching cost of $100,000 and an annual maintenance savings of $50,000. Assuming a discount rate of 3%, the discounted payback period would be longer than two years because the savings in year two would have an NPV of less than $50,000 (figure out the exact payback period for fun). But because we know that the there is a finite discounted payback period, we know that we should go ahead with the initiative.

7.7 Risk Management

What are software risks?

Software risks are "anything that can lead to results that deviate negatively from the stakeholders' real requirements for a project" [Gilb 2006].

How do risks manifest in software?

There are two kinds of software risks: external and internal. Internal risks include requirements changes, unrealistic requirements, incorrect requirements; shortfalls in externally furnished components; problems with legacy code; and lack of appropriate resources. These kinds of risks appear to be controlled most likely by the software project manager or her organization.

External risks are related to the business environment and include changes in the situation of customers, competitors, or suppliers; economic situations that change the cost structure; governmental regulations; weather, terrorism, and so on. Of course, the project manager can control none of these risks. Instead, she needs to plan for them so they can be mitigated when they arise.

How does the project manager identify, mitigate, and manage risks?

Many of the risks can be managed through close attention to the requirements specification and design processes. Prototyping (especially throwaway) is also an important tool in mitigating risk. Judicious and vigorous testing can reduce or eliminate many of these risks.

What are some other ways that the software project manager can mitigate risk?

Table 7.7, which is a variation of a set of recommendations from Boehm [1989], summarizes the risk factors and possible approaches to risk management and mitigation.

Is there a predictive model for the likelihood of any of these risks?

Yes. Once again, Boehm [1989] offers us some advice on the likelihood of various kinds of risks driving up cost, shown in Table 7.8.

TABLE 7.7

Various Project Risk Sources and Possible Management, Measurement, Elimination, and Mitigation Techniques

Risk Factor	Possible Management/Mitigation Approach
Incomplete and imprecise specifications	Prototyping Requirements reviews Formal methods
Difficulties in modeling highly complex systems	Prototyping Testing
Uncertainties in allocating functionality to software or hardware and subsequent turf battles	Prototyping Requirements reviews
Uncertainties in cost and resource estimation	Project management Metrics
Difficulties with progress monitoring	Project management Monitoring tools Metrics
Rapid changes in software technology and underlying hardware technology	Prototyping Testing
Measuring and predicting reliability of the software	Metrics Testing
Problems with interface definition	Prototyping
Problems encountered during software-software or hardware-software integration	Prototyping Testing
Unrealistic schedules and budgets	Project management Monitoring tools Metrics
Gold plating	Code audits and walkthroughs
Shortfalls in externally furnished components	Testing
Real-time performance shortfalls	Prototyping Testing
Trying to strain the limits of computer science capabilities	Code audits and walkthroughs Testing

Source: Adapted from Boehm, B.W., *Software Risk Management*, IEEE Computer Society Press, Los Alamitos, CA, 1989.

Do you have any other advice about management risk in software projects?

Tom Gilb, a software risk management specialist, suggests that the project manager ask the following questions throughout the life of the software project.

Why isn't the improvement quantified?

What is the degree of risk or uncertainty and why?

TABLE 7.8

Probabilistic Assessment of Risk

Cost Driver	Improbable (0.0–0.3)	Probable (0.4–0.6)	Frequent (0.7–1.0)
Application	Nonreal-time, little system interdependency	Embedded, some system interdependencies	Real-time embedded, strong interdependency
Availability	In place, meets need dates	Some compatibility with need dates	Nonexistent, does not meet need dates
Configuration management	Fully controlled	Some controls	No controls
Experience	High experience ratio	Average experience ratio	Low experience ratio
Facilities	Little or no modification	Some modifications, existent	Major modifications, nonexistent
Management environment	Strong personnel management approach	Good personnel management approach	Weak personnel management approach
Mix	Good mix of software disciplines	Some disciplines inappropriately represented	Some disciplines not represented
Requirements stability	Little or no change to established baseline	Some change in baseline expected	Rapidly changing or no baseline
Resource constraints	Little or no hardware-imposed constraints	Some hardware-imposed constraints	Significant hardware-imposed constraints
Rights	Compatible with maintenance and development plans	Partial compatibility with maintenance and development plans	Incompatible with maintenance and development plans
Size	Small, noncomplex, or easily decomposed	Medium to moderate complexity, decomposable	Significant hardware-imposed constraints
Technology	Mature, existent, in-house experience	Existent, some in-house experience	New or new application, little experience

Source: Boehm 1991.

Are you sure? If not, why not?

Where did you get that from? How can I check it out?

How does your idea measurably affect my goals and budgets?

Did we forget anything critical to survival?

How do you know it works that way? Did it before?

Do we have a complete solution? Are all requirements satisfied?

Are we planning to do the "profitable things" first?

Who is responsible for failure or success?

How can we be sure the plan is working during the project or earlier?

Is it "no cure, no pay" in a contract? Why not?

He offers other advice that reflects the healthy skepticism that the project manager needs to have:

- Re-think the deadline given — is it for real?
- Re-think the solution — is it incompatible with the deadline?
- What is the requestor's real need/point of view?
- Don't blindly accept "expert" opinions.
- Determine which components really must be delivered at the deadline [Gilb 2006].

How does prototyping mitigate risk?

Prototyping gives users a feel for how well the design approach works and increases communication between those who write requirements and the developers throughout the requirements specification and design process. Prototyping can be used to exercise novel hardware that may accompany an embedded system. Prototyping can also detect problems and identify deficiencies early in the life cycle, where changes are more easily and inexpensively made.

Are there risks to software prototyping?

Indeed, there are. For example, a prototype may not provide good information about timing characteristics and real-time performance, which lulls the designers into a false sense of security. Often the pressures of bringing a product to market lead to a temptation to carry over portions of the prototype into the final system. Therefore, use throwaway prototypes as much as possible.

Are there other ways to discover risks so that they can mitigated?

Yes. The best way is to ask experts who have worked on similar projects. There really is no substitute for experience.

7.8 Further Reading

Beck, K. et al., *The Agile Manifesto*, http://agilemanifesto.org/principles.html, accessed October 1, 2006.

Boehm, B., *Software Engineering Economics*, Prentice-Hall, Englewood Cliffs, NJ, 1981.

Boehm, B.W., *Software Risk Management*, IEEE Computer Society Press, Los Alamitos, CA, 1989.

Boehm, B. and Turner, R., *Balancing Agility and Discipline: A Guide to the Perplexed*, Addison-Wesley, Boston, MA, 2003.

Bramson, R., *Coping with Difficult People*, Dell Paperbacks, New York, 1988.

Brooks, F.P., *The Mythical Man Month, 20th Anniversary Edition*, Addison-Wesley, Boston, MA, 1995.

Cohen, C., Birkin, S., Garfield, M., and Webb, H., Managing conflict in software testing, *Commun. ACM*, 47(1), 76–81, 2004.

Covey, S.R., *Principle-Centered Leadership*, Simon & Schuster, New York, 1991.

Gilb, T., Principles of risk management,http://www.gilb.com/Download/Risk.pdf, accessed August 30, 2006.

IEEE 1490-1998 IEEE Guide — Adoption of PMI Standard — A Guide to the Project Management Body of Knowledge IEEE Standards, Piscataway, NJ, 1998.

Jones, C., *Patterns of Software Systems Failure and Success*, International Thomson Computer Press, Boston MA, 1996.

Laplante, P.A., *Software Engineering for Image Processing Systems*, CRC Press, Boca Raton, FL, 2003.

Laplante, P.A. and Neill, C.J., *Antipatterns: Identification, Refactoring, and Management*, Auerbach Publications, Boca Raton, FL, 2006.

Morgan, J.N., A roadmap of financial measures for IT project ROI, *IT Prof.*, Jan./Feb., 52–57, 2005.

Raffo, D., Settle, J., and Harrison, W., Investigating Financial Measures for Planning of Software IV&V, Portland State University Research Report #TR-99-05, Portland, OR, 1999.

Reifer, D.J., Web development: estimating quick-to-market software, *IEEE Software*, Nov./Dec., 57–64, 2000.

Reifer, D.J., Estimating Web development costs: there are differences, June 2002, available at www.stsc.hill.af.mil/crosstalk/2002/06/reifer.html, accessed 1/3/07.

Reifer, D., How good are agile methods?, *Software*, July/Aug., 16–18, 2002b.

Stelzer, D. and Mellis, W., Success factors of organizational change in software process improvement, *Software Process — Improvement and Practice*, 4, 227–250, 1998.

Thayer, R.H., Software system engineering: a tutorial, *Computer*, 35(4)68–73, 2002.

8

The Future of Software Engineering

Outline

- Open source
- Outsourcing and offshoring
- Globally distributed software development

8.1 Introduction

I would like to close this text with a discussion of recent changes in the software industry with a particular view of how these changes affect software engineering. First, we look at the important commercial issue of open source software (OSS) systems. Then we take up the often emotional issue of outsourcing* a software project, particularly to other countries (offshoring). Finally, we look at the very important issue of developing large software systems in multiple locations with distributed software teams. The issues raised in this new approach to software development are fascinating and complex.

8.2 Open Source

What is OSS?

OSS is software that is free use or free redistribution if the terms of the license agreement are followed. Usually this means that any work derived from the OSS can be redistributed only along with the source code and any derived works must comply with the same license.

* Much of Section 8.3 is excerpted from "The Who, What, Where, Why, and When of IT Outsourcing," by Phillip A. Laplante, Pawan Singh, Sudi Bindiganavile, Tom Costello, and Mark Landon, which appeared in *IT Professional*, January/February 2004, pp. 37–41. © 2004 IEEE, with permission.

TABLE 8.1

Various Kinds of OSS Projects

Type	Objective	Control Style	Community Structure	Major Problems	Examples
Exploration-oriented	Sharing innovation and knowledge	Cathedral-like Central control	Project leader Many readers	Subject to split	GNU systems JUN Perl
Utility-oriented	Satisfying an individual need	Bazaar-like Decentralized control	Many peripheral developers Peer support to passive users	Difficult to choose the right program	Linux system (excluding the kernel = exploration-oriented)
Service-oriented	Providing stable services	Council-like Central control	Core members instead of a project leader Many passive users that develop systems for end users	Less innovation	Apache PostgreSQL

Source: Nakakoji, K., Yamamoto, Y., Nishinaka, Y., Kishida, K., and Ye, Y., Evolution patterns of open-source software systems and communities, *Proc. Intl. Workshop on Principles of Software Evolution*, ACM Press, New York, May 2002.

OSS is the "opposite" of closed source software; that is, proprietary software for which the source cannot be had without paying a fee or had at all.

Where did OSS come from?

In 1983, Richard Stallman started the GNU project (which is a recursive acronym standing for GNU's Not Unix) to create a Unix-like operating system from free software. In 1991, Linus Torvalds created an operating system called Linux while a graduate student at the University of Helsinki.* Along the way, the process and culture created by Stallman and others, and carried on by Torvalds, formed the basis for the OSS movement today.

What kinds of code can be found as open source?

There are many thousands of open source projects ranging from games to programming languages, tools, and enterprise-level applications. Clones of many well-known desktop and enterprise applications are also available in open source. The different types of OSS projects are summarized in Table 8.1.

* Linux continues to be an important operating system found on many desktops today and Torvalds continues to direct the development of Linux. Other individuals, groups, and companies combine other components with Linux and redistribute it.

Some of the best-known projects are:

- Linux operating system
- Firefox Web browser
- Apache Web server
- GCC, Gnu C compiler

For the software engineer, there are many scripting languages like Perl, Python, PHP, and Ruby available in open source. Other useful developer tools include Ant and Maven for building applications; Hibernate, which acts as an object-oriented persistence layer for databases; XUnit for testing; CVS and SubVersion for source code control; and Eclipse or NetBeans as integrated development environments. I have discussed many of these tools in Chapter 5.

The software engineer may also want to use the open source Struts, which provides a framework in which an application can be built quickly using the model-view-controller architecture, and Swing, which provides a layered structure for managing business objects.

What is the value proposition for OSS?

The benefits of using OSS are clear — access to a large amount of sophisticated and useful code for free or nearly free, and for those applications with the most robust communities, it can be expected that the code will be maintained for many years.

OSS advocates also claim that OSS has fewer defects than closed source code. They quote Eric Raymond [1999] in this regard: "given enough eyeballs, all bugs are shallow." Of course, it is not true that OSS is error-free; some is quite buggy. But the research is still wide open on whether open source code is uniformly better, or worse, than closed source code. The truth is, it probably depends on the project.

What is the current state of OSS adoption?

OSS is being adopted more readily in Europe and South America than it is in the U.S. However, OSS adoption is picking up speed in the U.S.

OSS is probably already in your environment. But there is a great deal of misinformation causing delays in adoption and mistrust in some organizations. Therefore, in many situations, the opportunities and benefits may be under-realized or unrealized.

Who contributes to OSS systems?

Many people contribute code to open source repositories (or act as testers and beta users) for many different reasons. Individuals get involved to do something "important" or to be part of a community. Companies allow their employees to contribute code because the code has some benefit to the company. Others participate simply for fun or self-interest.

What different kinds of licenses are there?

The OSS licenses, and the community of OSS engineers, seek to preserve the integrity of all authors' work, not to discriminate against any persons, groups, or fields of endeavor.

There are roughly 100 types of licenses currently in use claiming to be open source variants. The Open Source Initiative (OSI) has approved 58 license types as being compliant with their stated criteria/goals.

But there are four "baseline" licenses originating from the late 1990s:

- General Public License (GNU GPL)
- GNU Lesser General Public License (LGPL)
- BSD License
- MIT License

I won't go into the differences between these licenses, however, as some of the differences are quite subtle and have important legal implications. Much more information on OSS licenses can be found at www.opensource.org/licenses/.

Where can I find OSSe projects?

OSS projects can be found in any number of open source repositories, the most popular of which is SourceForge (www.sourceforge.com). At this writing, there are more than 140,000 OSS projects in SourceForge.

Do companies really use OSS?

Most companies have some OSS in play, even if they decry its use. Many "open source forbidden" enterprises run Apache, use some of the open languages, or take advantage of open source Java libraries.

What are the characteristics of the OSS development model?

Characterization of the OSS development model is based on a model proposed by Eric Raymond [1999]. Raymond contrasts the traditional software development of a few people planning a cathedral in splendid isolation with the new collaborative "bazaar"* form of OSS development. OSS systems co-evolve with their developer communities, though the usual team critical mass of 5 to 15 core developers is needed to sustain a viable system.

What is software evolution?

Seminal work by Belady and Lehman [1976] in the 1970s involved studying 20 releases of the OS/360 operating system software. This was perhaps the first empirical study to focus on the dynamic behavior of a relatively large and mature 12-year-old system.

---

* Bazaar is a large number of developers coming together without monetary compensation to cooperate under a model of rigorous peer review and take advantage of parallel debugging that leads to innovation and rapid advancement in developing and evolving SW products.

TABLE 8.2

Lehman's Laws

No.	Brief Name	Lehman's Law
I 1974	Continuing Change	Software must be continually adapted or else progressively less satisfactory in use.
II 1974	Increasing Complexity	As software evolves, its complexity increases unless work is done to maintain or reduce it.
III 1974	Self-Regulation	Global software evolution processes are self-regulating.
IV 1978	Conservation of Organizational Stability	Unless feedback mechanisms are appropriately adjusted, the average effective global activity rate in an evolving software system tends to remain constant over product lifetime.
V 1978	Conservation of Familiarity	Generally, the incremental growth and long-term growth rate of software systems tend to decline.
VI 1991	Continuing Growth	The functional capability of software must continually increase to maintain user satisfaction over the system lifetime.
VII 1996	Declining Quality	The quality of software will decline unless it is rigorously adapted to accommodate changes in the operational environment.
VIII 1996	Feedback System (recognized 1971, formulated 1996)	Software evolution processes are multilevel, multiloop, multiagent feedback systems.

Source: Lehman, M.M., Perry, D.E., Ramil, J.F., Turski, W.M., and Wernick, P.D., Metrics and laws of software evolution — the nineties view, *Proc. Metrics '97*, IEEE–CS, Albuquerque, NM, November 5–7, 1997, pp. 20–32.

Belady and Lehman made a number of observations about the size and complexity of growth, which led them to postulate eight laws of software evolution dynamics (Table 8.2). Lehman's laws have important implications in both open and closed source development. For example, they might help explain why long-lived projects tend to follow the bathtub curve over time.

For example, a great deal of open source systems development takes the form of burst evolutions with periods of intensive rewrite. According to Lehman's Laws, the mean growth rate should be expected to decrease over time. However, larger OSS projects, such as the Linux kernel, have been shown to sustain super-linear growth [Godfrey and Tu 2000].

How does software requirements engineering occur in OSS?

Software requirements for OSS usually take the form of threaded messages and Web site discussions. Requirements engineering is closely tied to the interests of the OSS developer community [Scacchi 2004].

How do the software design and build processes take place in open source systems?

OSS code contributors (called committers) are often end users of the software. These committers usually employ versions of open source development tools such as CVS to empower a process of system build and incremental release review. Other members of the community use open source bug reporting tools (for example, Bugzilla) to dynamically debug the code and propose new features. Concurrent versions of systems play an essential role for coordination of decentralized code.

Software features are added evolutionary redevelopment, reinvention, and revitalization. The community of developers for each open source project generally experiences a high degree of code sharing, review, and modification, and redistributing concepts and techniques through minor improvements and mutations across many releases with short life cycles [Scacchi 2004].

How are OSS projects managed?

OSS project management consists of an organized interlinked layered meritocracy. This configuration is a hierarchical organizational form that centralizes and concentrates certain kinds of authority, trust, and respect for experience and accomplishment within the team. Such an organization is a highly adaptive but loosely coupled virtual enterprise [Scacchi 2004].

Are there downsides to using OSS?

Because of the dynamic nature of OSS, code versions and patches need to be carefully managed for any software that is adopted in the enterprise. There are companies, however, that manage these artifacts and provide "certified solution stacks" (collections of interoperable OSS) for a fee. Unless you contract such a company, however, the maintenance and security issues associated with using OSS applications could become a nightmare.

As far as using OSS in an end product, the licensing issues need to be well understood and managed. If they are not well understood, then the company may be at risk for legal troubles.

There is one other downside. Aside from the major OSS applications, which are highly sophisticated and developed, many of the software projects in open source repositories contain many bugs or lack important features because they are in various stages of maturity with developers of varying skill and motivation levels.

Despite the downsides, OSS or a blend of open source and new should be considered as part of any new software development initiative.

8.2.1 Software Archeology

What is software archeology?

Software archeology is a method of reconstructing the evolution of software using software trails. A software trail refers to any information created by developers of software, either explicitly or implicitly. Explicit trails include messages, code comments, and release notes. Implicit trails include information that can be extracted from the activity of software development itself such as release frequency, file size, file types, and number of LOC. This activity can be likened to an archeologist studying the remains of an ancient civilization in order to understand its history [German 2004].

The study of software evolution has been enabled by the advent of OSS. With OSS, researchers are able to access the software trails of thousands of projects. In contrast, proprietary, closed source projects are unlikely candidates for study because organizations have a clear business case for investing in such activity and are not likely to release their trails, even if divorced from the source code.

What is software archeology used for?

Aside from purely academic reasons for studying OSS, software archeology is important to those software engineering groups that are considering adopting or becoming involved in an open source project. Potential adopters can use software archeology to study the quality of the candidate application (for example, how closely it follows Lehman's Laws). Potential participants in the project can use archeological techniques to study the history and context of the software and community.

Software archeological techniques can also be used to study closed source software if access to the relevant artifacts is available.

Can you give an example of an archeological study?

An example is a nice way to summarize much of what I have discussed throughout this text.* The subject of this study is jEdit, which was designed to be a text editor to aid in software development. As such, it provides features that help in writing, analyzing, and debugging code.

One of the easiest software trails to uncover is the development over time of versions of jEdit. This software trail was found at the SourceForge Web site (http://sourceforge.net/projects/jedit) where jEdit development is currently being hosted. This Web site provides a list of major releases of jEdit available for download along with the dates they occurred. The history of the major releases of jEdit is shown in Figure 8.1.

Another software trail that provided information about the development of jEdit is the developer e-mail archives found at SourceForge. These e-mails contained developer discussions that chronicled the developer activity during the evolution of jEdit. A chart depicting the email activity that occurred in the developer mailing list from December 1999 to April 2006 is in Figure 8.2. This e-mail traffic tells a story about occurrences within the developer

* Thanks to Ken Sassa for conducting and writing this archeological study.

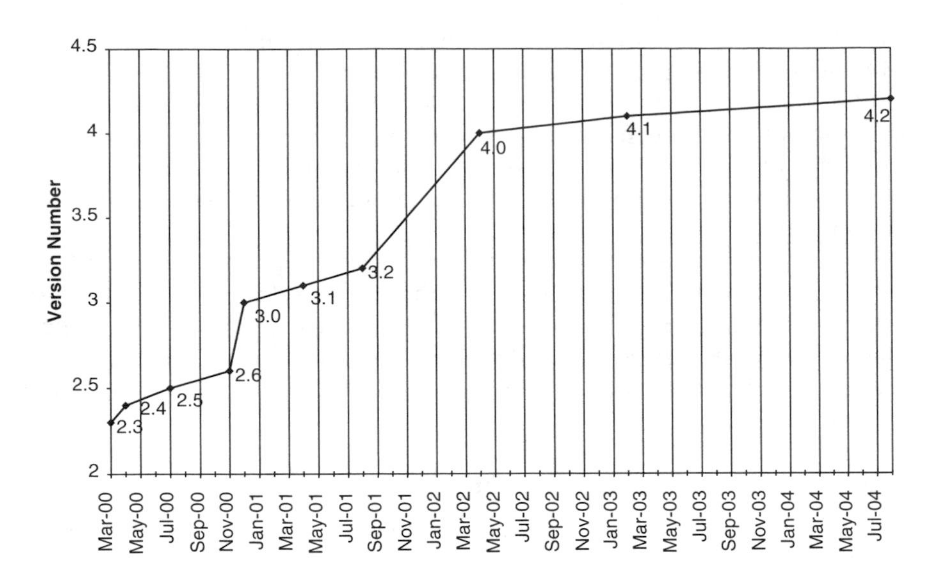

FIGURE 8.1
Major releases of jEdit.

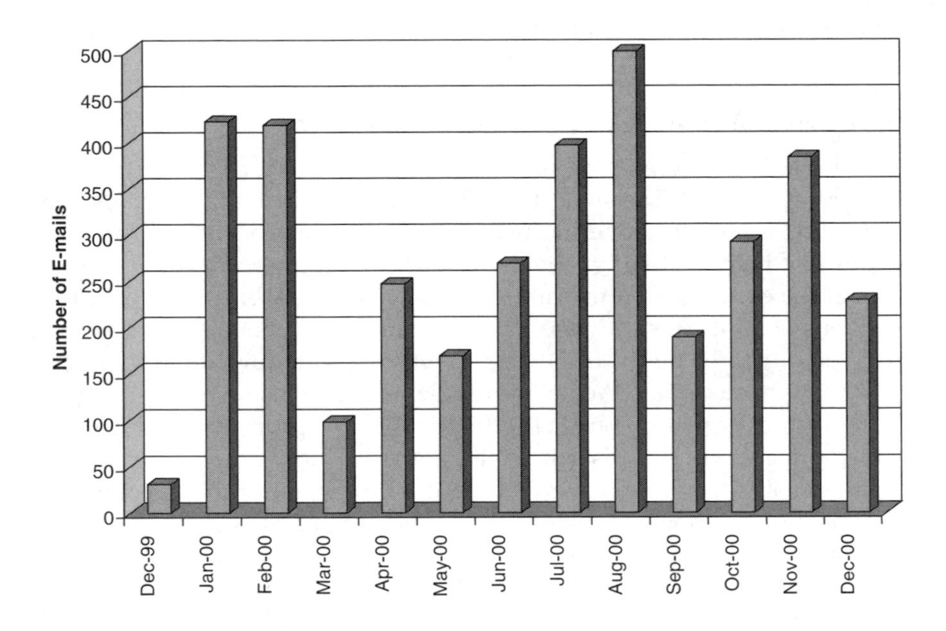

FIGURE 8.2
Developer mailing list activity (December 1999 to December 2000) for jEdit.

community, a story that is beyond the scope of this text but may be of interest to jEdit users.

A further software trail that gave insight into the evolution of jEdit was the number of opened and closed bugs that were reported. Tracking the number of bugs opened over time proved quite easy to do because the open date was given as a column heading in the bug tracking system for jEdit at SourceForge. Tracking the number of closed bugs proved a bit more difficult because the closed date was not easily accessible. Rather, the closed date was buried in each individual bug report. Thus, getting the closed date required opening each bug report, getting the closed date for the bug, and putting a mark in the month and year for that closed bug. Then the number of marks was counted for each individual month and this number represented the bugs closed for that month. This was done for all the bugs from July 2000 through December 2002. In this manner, the bugs closed vs. bugs opened could be tracked. The result of this process is shown in Figure 8.3.

Finally, we can look at the evolution of jEdit by focusing on the development of four versions of jEdit. The versions chosen for this examination are four major jEdit releases: 3.0, 3.2, 4.0, and 4.2. The development of these four versions can be tracked from five different perspectives, which include development in terms of LOC, graph cycles, cyclomatic complexity, cohesion and coupling, and code smells. The software trails that are used to analyze the development of these four jEdit versions are the source code that is available at SourceForge for these jEdit releases.

In order to analyze the source code and study its development from the varied perspectives, a source code visualization and analysis tool called

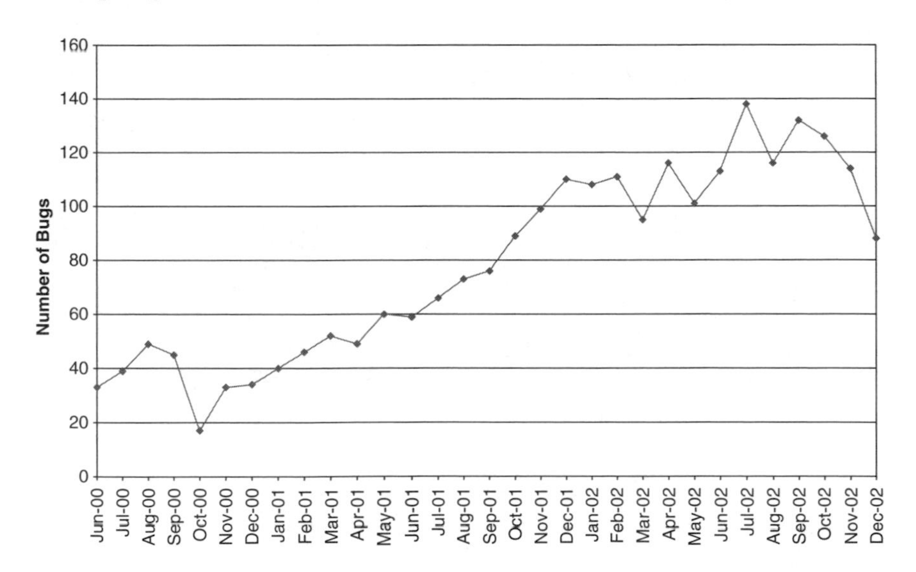

FIGURE 8.3
Number of bugs reported over time in SourceForge for jEdit from June 2000 through December 2002.

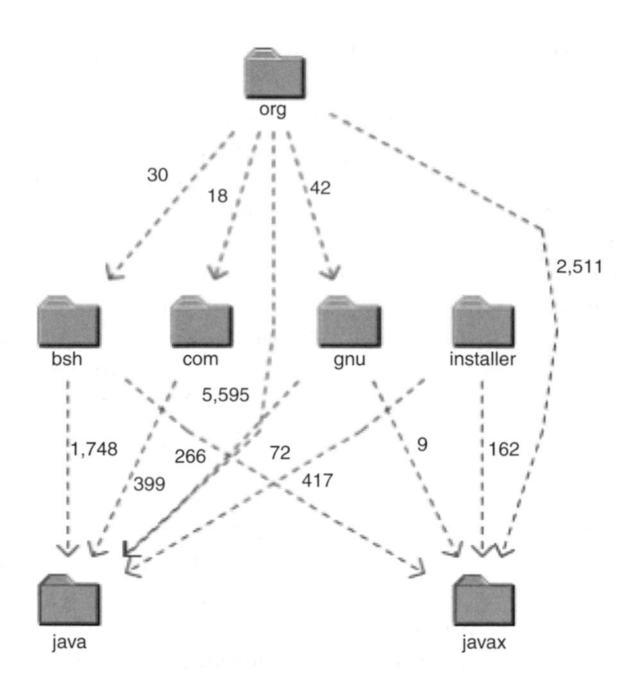

FIGURE 8.4
jEdit 3.0 higraph.

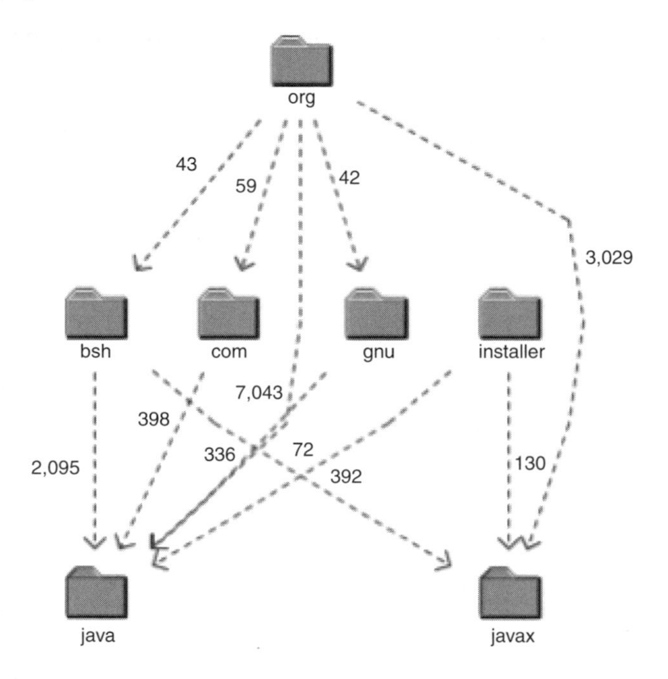

FIGURE 8.5
jEdit 3.2 higraph.

Headway reView was employed. This tool has the capability to count lines of code; find graph cycles; measure cyclomatic complexity, cohesion, and coupling; and search the code for code smells. Analysis begins with version 3.0 because no source code was available for versions earlier than 3.0. Such source code is needed in order for Headway re View to analyze the code. SourceForge does have installation JAR files for versions prior to 3.0, but no source code for them that can be compiled and fed to Headway reView.

Before looking at the development of the four versions in detail, it is helpful to look at an overview of them first. Such an overview is found in the higraphs (a visualization of the software package diagrams) that Headway reView generated for jEdit which are presented in Figure 8.4 through Figure 8.7.

In the figures, the names of the packages are labeled on the folder icon. The five folders that make up the structure of jEdit are `bsh`, `com`, `gnu`, `org`,

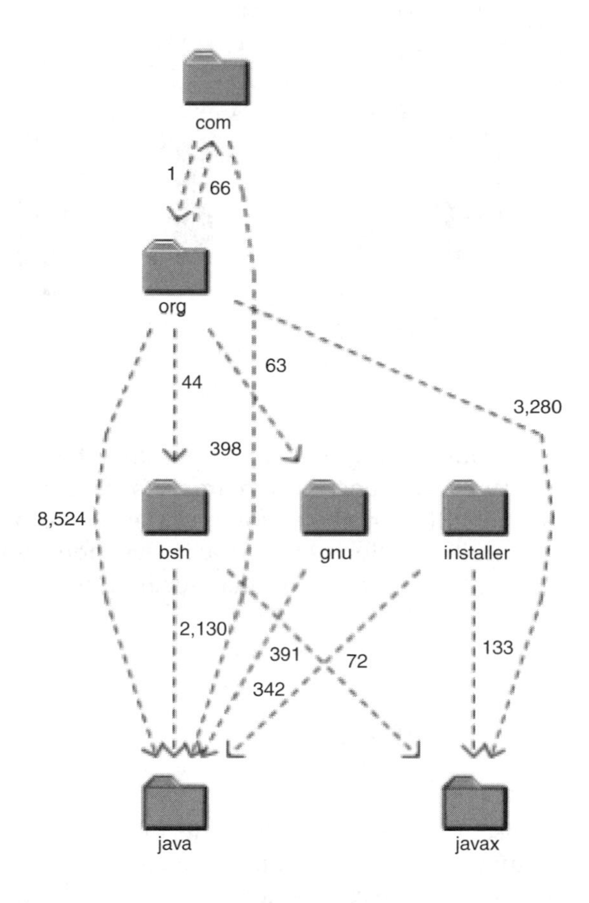

FIGURE 8.6
jEdit 4.0 higraph.

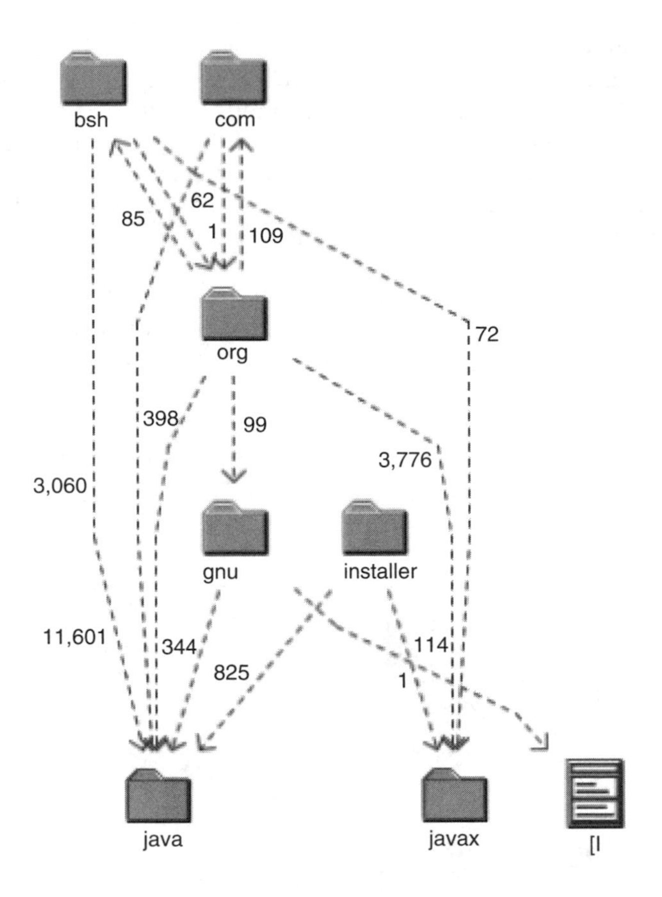

FIGURE 8.7
jEdit 4.2 higraph.

and `installer`. The number on the edges represents the number of times the subordinate package is referenced. More importantly, notice that in some cases cycles developed in the structures; for example, between `com` and `org` in Figure 8.6. Cycles are a violation of the Acyclic Dependency Principle and indicate potential difficulties in program maintenance and understanding.

8.3 Outsourcing and Offshoring

What is outsourcing?

Outsourcing is the use of outside firms to perform activities that were previously performed internally. Classic examples include electrical component manufacturing and perhaps the earliest form of information processing outsourcing — payroll processing.

For the software engineering industry, outsourcing means subcontracting some or all of the software development life-cycle processes. Companies can outsource software requirements engineering, design, development, testing, documentation, and even project management. Outsourcing can be done domestically or to another country, as is increasingly common today. In the latter case, it is called offshoring. Hiring a few consultants to serve as internal members of a software engineering team, however, is not considered outsourcing.

Why do companies outsource?

Outsourcing is generally done for cost savings, to achieve a better focus on the core business, or because certain software engineering functions are considered to be inefficiently or ineffectively handled internally. From an economic standpoint, if any software engineering activity can be considered a commodity, then there is little justification for performing that activity internally. In these cases, a focused vendor should be able to provide the service at a higher level of quality, lower cost, or both. In other words, outsourcing takes advantage of the economies of scale of another business that specializes in that domain.

Which organizations should outsource?

Perhaps it is easier to answer the question, "Which organizations should not outsource?" There is a myth that outsourcing is cheap. It is not. Even in India, where the perceived difference in relative economy would suggest a lower labor cost, the cost of a skilled developer is approximately $40 per hour at this writing. In most cases, vendors will only take on large projects — making outsourcing less desirable for small software projects.

But in all cases, a strong communication infrastructure is needed to make outsourcing work. The infrastructure costs could include significant domestic and international travel, telecommunications costs, providing specialized equipment to the vendor, and so on. Therefore, it is easy to conclude that outsourcing is generally not for very small organizations.

To whom should you outsource?

In answering this question, consider that you are transferring knowledge when you outsource. This knowledge can be valuable. It is possible that a vendor can cut-and-run after the outsourcing project is completed, and nondisclosure or noncompete agreements are more difficult to prosecute if the outsourced vendor is not based in the U.S. Therefore, the vendor must be a trusted one.

In choosing a vendor, you must also be very careful about protecting your brand through accountability for the actions of the vendor and by transference of the reputation of the vendor. Remember that the vendors to whom you outsource should view your company as a partner, not as a client to be milked.

What are some issues involved in offshoring?

Offshoring is becoming increasingly popular. Australia, India, Ireland, New Zealand, the former Soviet Bloc countries (such as Bulgaria and Russia), Brazil, Argentina, and Venezuela, to name just a few, have become favorite destinations for American projects.

When dealing with vendors overseas, investigate a number of issues such as competence, reliability, and quality. It could be disastrous to discover that an outsourced vendor has provided inferior software components. For this reason, many companies insist on indications of quality — for example, that the vendor is at CMM Level 4 and higher — before even considering a relationship.

Pay attention to the legal organization of the potential vendor. For example, a vendor that is organized as a subsidiary of another company can be disconnected in the event of a lawsuit, making it difficult to obtain remedies in the case of malpractice. Similarly, disputes with an overseas vendor might be more difficult to resolve because of cultural differences and because legal remedies are more complicated and costly to obtain.

Whether the outsourcing is domestic or foreign, the chemistry and culture of the vendor and client have to mix.

Where in the enterprise should outsourcing be used?

A commonly held view on where to use outsourcing is one that is similar to the economic notion that the market does not pay a company for diversifying risk. Analogously, the market does not pay a company for doing things outside its core business. A straightforward interpretation of this statement is that any function of a business that is not part of its core is a candidate for outsourcing.

Any software engineering function could be considered for outsourcing; for example, requirements engineering, software architecture, software design, code development, testing, documentation, and project management

When should a company not outsource?

In cases where the benefit is low, there is no need to outsource. In cases where the benefit is high and the function is not within the core business, there is a strong incentive to outsource. However, when the function is outside of core business but the potential outsourcing benefit is low or when the potential outsourcing benefit is high but the function is within the core business, it is probably not work the risk to outsource.

When should outsourcing be done and at what stage of the process?

You usually recognize the need to outsource when it is too late. Therefore, the decision to outsource must be made early, and you must analyze the cost–benefit ratio of waiting too long to decide.

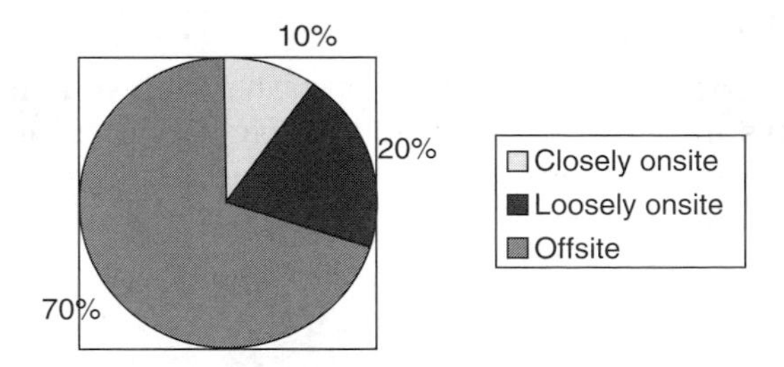

FIGURE 8.8
A three-tier strategy for outsourcing.

How do companies outsource?

There are several standard methodologies for outsourced software development. For example, consider the three-tier model (Figure 8.8). Figure 8.8 depicts a recommended distribution of outsourced work in which 10% of the work is held closely and performed onsite by the vendor's staff under close supervision. Another 20% of the effort is also done onsite by a combination of vendor and in-house staff under normal supervision. The remaining 70% of the project is done entirely offsite by the vendor.

When outsourced work is performed offsite, it is critical that your own agent is at the site of the outsourcing. A good rule of thumb is to have one of your staff present for every 20 staff working offsite. The rule of 20 is based largely on the incremental cost of housing an agent offshore to supervise the work.

Outsourcing can be a learning endeavor. It might seem rather mercenary to bring in a vendor to learn from and then jettison, but this is a risk vendors understand and factor into their margins.

Do you have any rules of thumb for outsourcing and offshoring?

The following best practices come to mind:

Whether the project is outsourced domestically or overseas, have your own agent at the vendor's site: 1 agent for every 20 persons outsourced.

Make sure that you understand the true cost structure of outsourcing and offshoring. There may be many unaccounted for hidden costs.

When negotiating the contract and throughout the project life cycle, carefully set expectations.

Have a quality management infrastructure in place.

In the case of overseas outsourcing, take into account language, culture, and time differences.

Make sure up front that there is a detailed process of project definition and specification development. This ensures that the project methodology, scope, schedule, and deliverables are unambiguously defined and understood by both parties.

Does outsourcing ever fail?

Many companies fail in the execution of strategic outsourcing. There are a number of ways in which they can fail: organization culture mismatches, morale damage due to layoffs, or outsourcing for the wrong reasons.

Outsourcing cannot absolve you of your responsibility. You can't outsource your problems.

8.4 Global Software Development

What is global software development?

Global software development (GSD) is software development in a multisite, multicultural, and distributed context. In GSD, software development is viewed as a geographically distributed network of workflows. A central team at a single location typically designs requirements and high-level design. But subsystems designed by the central team are developed by remote teams at multiple different locations

What is the business case for GSD?

First, GSD can take advantage of lower development costs in other countries. Moreover, if a limited trained workforce is available in a single location in certain technologies, it may be found elsewhere. The offshoring aspect is perhaps the most controversial because of the political risks of sending jobs to other countries. However, this consideration is beyond the scope of this book.

The time zone differences between teams, which could be considered as a negative when it comes to trying to conduct teleconferences, can be an advantage if a shift-based system facilitated by time-zone differences is used. For example, if three teams are distributed in time zones that differ by 8 hours, then one team can write code, the second team can debug, integrate, and build an executable, and the third team can test the build. This allows for single increments to be completed in 24 hours. Advances in telecommunications infrastructure and special collaborative software tools have made this kind of distributed development possible.

What software process can be used for GSD?

The answer to this question presents somewhat of a paradox. One would think that whatever process is used that it should be lightweight to minimize communication overhead and allow for easy measurement of progress based on a working product rather than completed documents. It is probably necessary that the process model itself be adaptable. Such criteria suggest that an agile process model should be used.

But agility seems counterintuitive in a distributed development environment because agility requires close collaboration. Therefore, collocated teams would be ideal. Similarly, more, not less, documentation would seem to be needed to unambiguously communicate information to remote teams.

Therefore, we likely need a process model that enjoys some features of the traditional models and some features of agile methodologies.

It turns out that the standard UPM is a good place to start. Recall that the UPM is iterative and incremental, use case driven, architecture centric, and actively manages risks. We can map the traditional four phases of the UPM into activities that have significant milestones for the GSD, namely:

Inception: software vision is communicated

Elaboration: software architecture is developed

Construction: software is built

Transition: product is released and tested

As with the UPM, these phases can be iterated.

Certain agile practices can be followed throughout the product development life cycle. For example, short time iterations test first development, continuous refactoring, pair programming and stand up meetings can all be facilitated using Web casting, instant messaging, or even e-mail and telephone. Other aspects such as collective code ownership and onsite customers can be challenging. Most aspects of any agile development methodology (e.g., XP or SCRUM), however, can be easily supported during the construction phase.

Due to the limitation of certain agile practices, to scale up to large distributed projects it becomes important to maintain comprehensive requirements and design models. User stories and back-of-the-napkin design will not do. Some of this burden can be alleviated by automating generation of some documents such as an SRS and SDS using the UML model and acceptance tests.

What are the challenges for GSD?

For any project to be successful, whether collocated or distributed, a process framework needs to orchestrate communication, coordination, and control effectively. For example, teams working on a given project need to communicate with each other and their activities need to be synchronized and controlled to enforce common goals, policies, standards, and quality levels.

But when the teams working on a project are geographically distributed, communication, coordination, and control become challenging. Distance exacerbates the problem of clearly and unambiguously communicating information between collaborating teams. This aspect becomes more complex if teams are separated by several time zones. Distance also makes it difficult to coordinate tasks within a distributed development environment. The distributed development environment makes it difficult to enforce common goals, policies, standards, and quality levels among collaborating teams.

Another key aspect of successful projects is maintaining good team dynamics. Team building can be challenging in distributed projects because members may not know each other, language and culture may be different, and the daily processes to be followed by members at different sites may vary.

How can these challenges be overcome?

One way to overcome the challenges of communication is by optimizing the collaboration among the distributed teams. Collaboration can be optimized by breaking a system into loosely coupled subsystems or components and having each team work on a given subsystem or component independent of each other.

Team management problems can be overcome by assigning project members who straddle multiple teams. These individuals act as ambassadors or liaisons between teams and help bridge the communication gap. These people must be engaged early in the inception and elaboration phase as this helps them to acquire the knowledge of the domain and gain an understanding of the architecture. With the acquired knowledge, they can share the architectural vision and help answer a majority of the architecture- and domain-related questions from the development teams.

8.5 Further Reading

Belady, L.A. and Lehman, M.M., A model of large program development, *IBM Syst. J.*, 15(1), 225–252, 1976.

Damian, D. and Moitra, D., Global software development: How far have we come? *IEEE Software*, 23(5), 17–19, 2006.

German, D., Using software trails to reconstruct the evolution of software, *J. Software Mainten. Evol.: Res. Practice*, 16(6), 367–384, 2004.

Godfrey, M. and Tu, Q., Evolution in open source software: A case study, *Proc. 2000 Intl. Conf. Software*

Mainten. (ICSM 2000), San Jose, CA, October 2000.

Johnson, R.E. Reverse engineering and software archaeology, *The DoD SoftwareTech News*, 8(3), 4–8, October 2005.

Koru, A. and Tian, J., Comparing high-change modules and modules with the highest measurement values in two large-scale open-source products, *IEEE Trans. Software Eng.*, 31(8), 625–642, 2005.

Laplante, P.A., Singh, P., Bindiganavile, S., Costello, T., and Landon, M., The who, what, where, why, and when of IT outsourcing, *IT Prof.*, 6(1) 37–41, 2004.

Lehman, M.M., Perry, D.E., Ramil, J.F., Turski, W.M., and Wernick, P.D., Metrics and laws of software evolution — the nineties view, *Proc. Metrics '97*, IEEE–CS, Albuquerque, NM, November 5–7, 1997, pp. 20–32.

Nakakoji, K., Yamamoto, Y., Nishinaka, Y., Kishida, K., and Ye, Y., Evolution patterns of open-source software systems and communities, *Proc. Intl. Workshop on Principles of Software Evolution*, ACM Press, Los Alamitos, CA, May 2002.

Paulson, J.W., Succi, G., and Eberlein, A., An empirical study of open-source and closed source software products, *IEEE Trans. Software Eng.*, 30(4), 246–256, 2004.

Raymond, E. S., *The Cathedral and the Bazaar*, O'Reilly, Cambridge, MA, 1999.

Sangwan, R., Neill, C., Laplante, P., Paulish, D., and Kuhn, W., A framework for agile development in outsourced environments, *WSEAS Trans. Comput.*, 3(5), 1530–1537, 2004.

Sangwan, R., Bass, M., Mullick, N., and Paulish, D., *Managing Global Software Development*, Auerbach Publications, Boca Raton, FL, 2006.

Scacchi, W., Free and open source development practices in the game community, *IEEE Software*, 21(1), 59–66, 2004.

Schneider, A. and Windle, P., Software archaeology, *The DoD SoftwareTech News*, 8(3), 9–13, October 2005.

Appendix A

Software Requirements for a Wastewater Pumping Station Wet Well Control System (rev. 01.01.00)

Christopher M. Garrell

A.1 Introduction

A wastewater pumping station is a component of the sanitary sewage collection system that transfers domestic sewage to a wastewater treatment facility for processing. A typical pumping station includes three components: (1) a sewage grinder, (2) a wet well, and (3) a valve vault (Figure A.1). Unprocessed sewage enters the sewage grinder unit so that solids suspended in the liquid can be reduced in size by a central cutting stack. The processed liquid then proceeds to the wet well, which serves as a reservoir for submersible pumps. These pumps then add the required energy/head to the liquid so that it can be conveyed to a wastewater treatment facility for primary and secondary treatment. The control system specification that follows describes the operation of the wet well.

A.1.1 Purpose

This specification describes the software design requirements for the wet well control system of a wastewater pumping station. It is intended that this specification provide the basis of the software development process and be used as preliminary documentation for end users.

A.1.2 Scope

The software system described in this specification is part of a control system for the wet well of a wastewater pumping station. The control

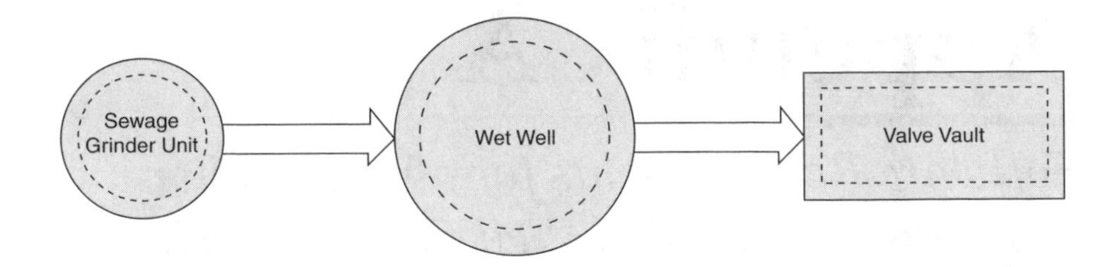

FIGURE A.1
Typical wastewater pumping station process.

system supports an array of sensors and switches that monitor and control the operation of the wet well. The design of the wet well control system shall provide for the safety and protection of pumping station operators, maintenance personnel, and the public from hazards that may result from its operation. The control system shall be responsible for the following operations:

a. Monitoring and reporting the level of liquid in the wet well.
b. Monitoring and reporting the level of hazardous methane gas.
c. Monitoring and reporting the state of each pump and noting whether it is currently running or not.
d. Activating a visual and audible alarm when a hazardous condition exists.
e. Switching each submersible pump on or off in a timely fashion depending on the level of liquid within the wet well.
f. Switching ventilation fans on or off in a timely fashion depending on the concentration of hazardous gas within the wet well.

Any requirements that are incomplete are annotated with "TBD" and will be completed in a later revision of this specification.

A.1.3 Definitions, Acronyms, and Abbreviations

The following is a list of definitions for terms used in this document.

Audible alarm — The horn that sounds when an alarm condition occurs.

Controller — Equipment or a program within a control system that responds to changes in a measured value by initiating an action to affect that value

DEP — Department of Environmental Protection.

Detention basin — A storage site, such as a small, unregulated reservoir, which delays the conveyance of wastewater.

Effluent — Any material that flows outward from something; an example is wastewater from treatment plants.

EPA — Environmental Protection Agency.

Influent — Any material that flows inward from something; an example is wastewater into treatment plants.

Imminent threat — A situation with the potential to immediately and adversely affect or treaten public health or safety.

Manhole — Hole, with removable cover, through which a person can enter into a sewer, conduit, or tunnel to repair or inspect.

Methane — A gas formed naturally by the decomposition of organic matter.

Overflow — An occurrence by which a surplus of liquid exceeds the limit or capacity of the well.

Pre-cast — A concrete unit that is cast and cured in an area other than its final position or place.

Pump — A mechanical device that transports fluid by pressure or suction.

Remote override — A software interface that allows remote administrative control of the pumping control system.

Seal — A device mounted in the pump housing and/or on the pump haft, to prevent leakage of liquid from the pump.

Security — Means used to protect against unauthorized access or dangerous conditions. A resultant visual and/or audible alarm is then triggered.

Sensor — The part of a measuring instrument that responds directly to changes in the environment.

Sewage grinder — A mechanism that captures, grinds, and removes solids, ensuring a uniform particle size to protect pumps from clogging.

Submersible pump — A pump having a sealed motor that is submerged in the fluid to be pumped.

Thermal overload — A state in which measured temperatures have exceeded a maximum allowable design value.

Valve — A mechanical device for controlling the flow of a fluid.

Ventilation — The process of supplying or removing air by natural or mechanical means to or from a space.

Voltage — Electrical potential or electromotive force expressed in volts.

Visible alarm — The strobe light that is enabled when an alarm condition occurs.

Wet well — A tank or separate compartment following the sewage grinder that serves as a reservoir for the submersible pump.

A.2 Overall Description

A.2.1 Wet Well Overview

The wet well for which this specification is intended is shown in Figure A.2. The characteristics of the wet well described in this specification are as follows.

a. The wet well reservoir contains two submersible pumps sized to provide a fixed capacity.
b. Hazardous concentrations of flammable gases and vapors can exist in the wet well.
c. It has a ventilation fan that is oriented to direct fresh air into the wet well rather than just remove exhaust from the well.
d. An alarm and indicator light are located outside so that operators can determine if a hazardous condition exists. Hazardous conditions include, but are not necessarily limited to, a high gas level, a high water level, and pump malfunction.
e. A float switch is used to determine the depth of liquid currently in the wet well.

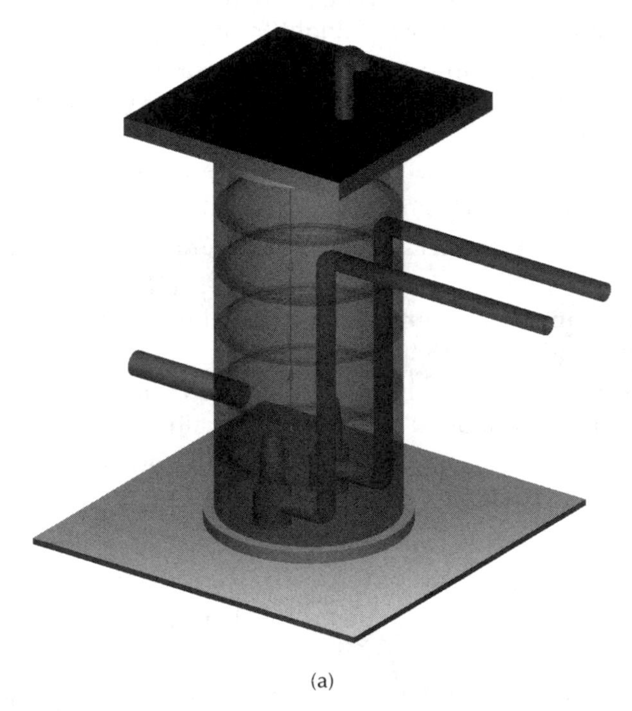

(a)

FIGURE A.2
Typical wet well. (a) Three-dimensional view. (b) Top view schematic. (c) Side sectional view.

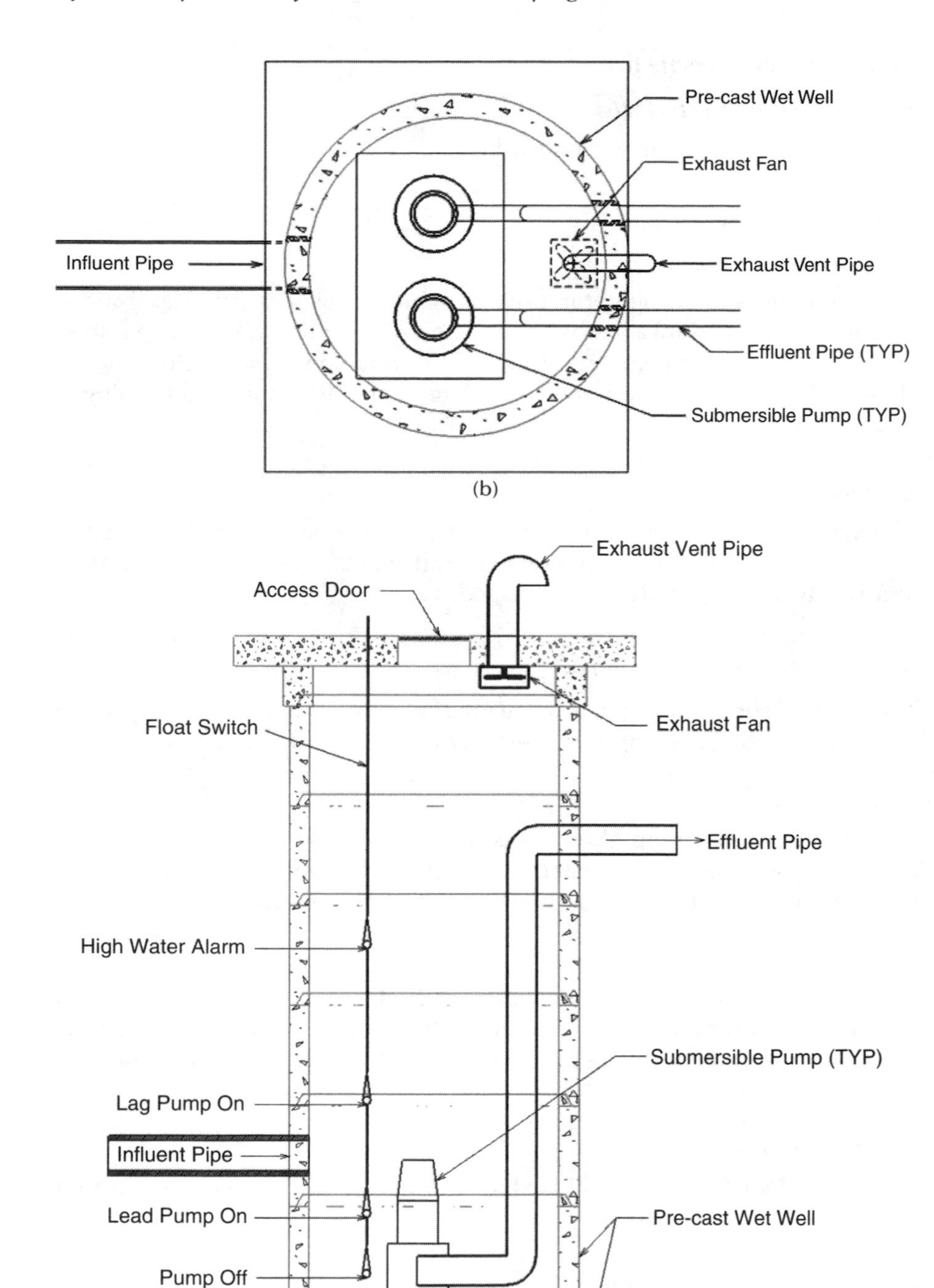

FIGURE A.2
(Continued).

A.2.2 Product Perspective

A.2.2.1 System Interfaces

The system interfaces are described below.

A.2.2.2 User Interfaces

Pumping Station Operator

The pumping station operator uses the control display panel and alarm display panel to control and observe the operation of the submersible pumps and wet well environmental conditions. Manipulation of parameters and the state of the submersible pumps is available when the system is running in manual mode.

Maintenance Personnel

The maintenance personnel use the control display panel and alarm display panel to observe the current parameters and state of the submersible pumps and wet well and perform maintenance.

A.2.2.3 Hardware Interfaces

The wet well control system hardware interfaces are summarized in Figure A.3. The major hardware components are summarized in Table A.1.

Moisture Senor

Each submersible pump shall be equipped with a moisture sensor that detects the occurrence of an external pump seal failure. Should a seal failure be detected, the pump shall be turned off and alarm state set.

Float Switch

The float switch is a mercury switch used to determine the depth of liquid within the wet well and set the on or off state for each pump. Three switch states have been identified as lead pump on/off, lag pump on/off, and high water alarm.

Access Door Sensor

The access door sensor is used to determine the state, either opened or closed, of the wet well access door.

A.2.2.4 Software Interfaces

Pump Control Unit

The wet well control system interfaces with the pump control system, providing a pump station operator and maintenance personnel with the ability to observe the operation of the submersible pumps and wet well environmental conditions. The pump control unit provides the additional capability

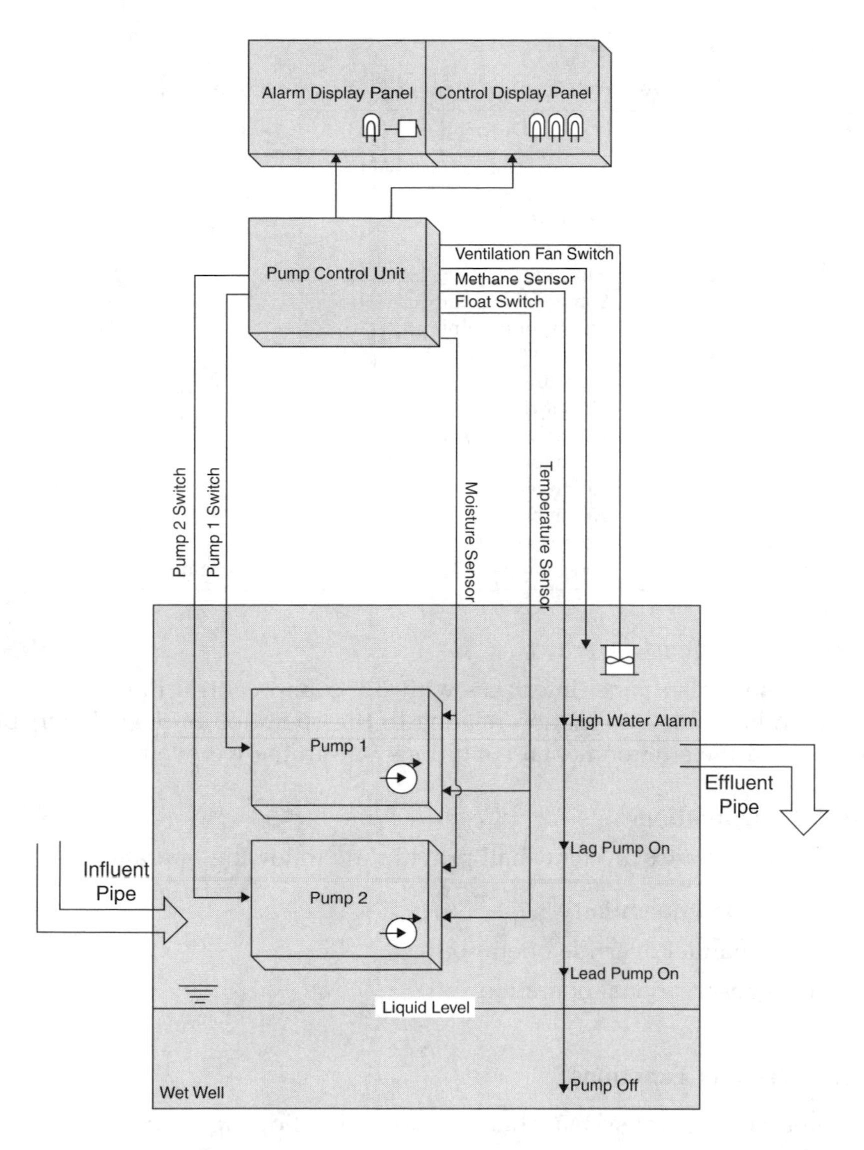

FIGURE A.3
Wet well control system hardware.

of manipulation of parameters and states of the submersible pumps when the system is running in manual mode.

Control Display Panel

The control display panel interfaces with the pump control unit, providing visual information relating to the operation of the submersible pumps and environmental conditions within the wet well.

TABLE A.1

Major Wet Well Control System Hardware Components

Item	Description	Quantity
1	Pre-cast concrete wet well	1
2	Access door	1
3	Ventilation pipe	2
4	Axial flow fan	2
4.1	Fan switch	2
5	Submersible pump	2
6	Pump control unit	1
6.1	Temperature sensor	2
6.2	Moisture sensor	2
6.3	Float switch	1
6.4	Access door sensor	1
7	Alarm panel	1
7.1	Alarm lamp	1
7.2	Alarm buzzer	1
8	Control panel	1
8.1	Panel lamps	6 (3 per pump)

Alarm Display Panel

The alarm display panel interfaces with the pump control unit, providing visual and audible information relating to the operation of the submersible pumps and the environmental conditions within the wet well.

A.2.2.5 Operations

The wet well control system shall provide the following operations:

a. Automated operation
b. Local manual override operation
c. Local observational operation

A.2.3 Product Functions

The wet well control system shall provide the following functionality.

a. Start the pump motors to prevent the wet well from running over and stop the pump motors before the wet well runs dry.
b. Keep track of whether or not each motor is running.
c. Monitor the pumping site for unauthorized entry or trespass.
d. Monitor the environmental conditions within the wet well.
e. Monitor the physical condition of each pump for the existence of moisture and excessive temperatures.
f. Display real-time and historical operational parameters.
g. Provide an alarm feature.

h. Provide a manual override of the site.
i. Provide automated operation of the site.
j. Equalize the run time between the pumps.

A.2.4 User Characteristics

Pumping Station Operator — Authorized personnel trained with the usage of the wet well control system when it is in manual mode.

Maintenance Personnel — Authorized personnel trained with the usage of the wet well control system.

A.2.5 Constraints

System constraints include the following items:

a. Regulatory agencies including but not limited to the EPA and DEP.
b. Hardware limitations.
c. Interfaces to other applications.
d. Security considerations.
e. Safety considerations.

A.2.6 Assumptions and Dependencies

Assumptions and dependencies for the wet well control system include the following items:

a. The operation of the sewage grinder unit is within expected tolerances and constraints at all times.
b. A power backup system has been proved as a separate system external to the wet well control system.
c. The operation of the controls within the valve vault is within expected tolerances at all times.

A.3 Specific Requirements

The following section defines the basic functionality of the wet well control system.

A.3.1 External Interface Requirements

See Figure A.4 for the use case diagram.

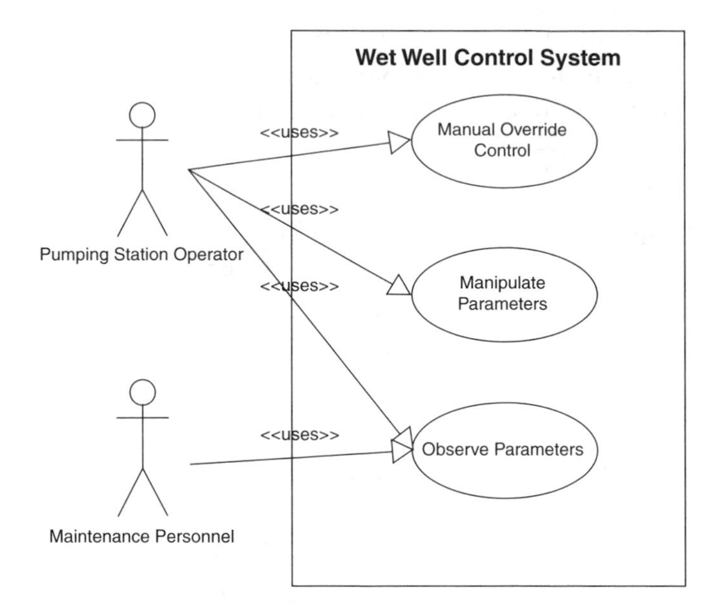

FIGURE A.4
Use case diagram.

A.3.2 Classes/Objects

A.3.2.1 Pump Control Unit

The pump control unit shall start the submersible pump motors to prevent the wet well from running over and stop the pump motors before the wet well runs dry within 5 seconds. LeadDepth represents the depth of liquid when the first pump should be turned on. LagDepth represents the depth of liquid when the second pump should be turned on. HighDepth represents the depth of liquid that the wet well should be kept below. Should the depth of liquid be equal to or exceed HighDepth, the alarm state is set. AlarmState represents a Boolean quantity such that at any time t, the audible and visual alarms are enabled. Depth represents the amount of liquid in the wet well at any time t in units of length. Pumping represents a Boolean quantity such that at any time t, the pumps are either running or not.

$$Depth : Time \rightarrow Length$$

$$Pumping : Time \rightarrow Bool$$

$$HighDepth > LagDepth > LeadDepth$$

$$Depth \geq LagDepth \Rightarrow Pumping$$

$$Depth \geq HighDepth \Rightarrow AlarmState$$

a. The pump control unit shall start the ventilation fans in the wet well to prevent the introduction of methane into the wet well within 5 seconds of detecting a high methane level.

b. The pump control unit shall keep track of whether or not each motor is running.

c. The pump control unit shall keep track of whether or not each motor is available to run.

d. If a pump motor is not available to run and a request has been made for the pump motor to start, an alternative motor should be started in its place.

e. An alarm state shall be set when the high water level is reached.

f. An alarm state shall be set when the high methane level is reached.

g. The starting and stopping of the pump motors shall be done in manner that equalizes the run times on the motors.

h. Level switches shall be used to indicate when pump motors should be started.

i. The pump control unit shall be notified if excess moisture is detected in a pump motor.

j. The pump control unit shall be notified if a pump motor overheats and shall shut down the overheated motor.

k. The pump control unit shall be responsible for monitoring the pumping site.

l. The pump control unit shall be responsible for recording real-time and historical operational parameters.

m. The pump control unit shall be responsible for providing an alarm feature.

n. There shall be an automatic and manual mode for the pump control unit. Each pumping station shall be in either automatic mode or manual mode.

o. Monitor and detect prohibited entry to the wet well through the access door by way of a broken electrical circuit. Both audible and visible alarms are activated.

p. Monitor and detect occurrence of a pump motor seal leak. If a leak has been detected, both an audible and visible alarm should be activated within 5 seconds.

A.3.2.2 Control Display Panel

a. The control display panel shall have a digital depth of influent measured in feet.

b. Monitor and detect prohibited entry by way of a broken electrical circuit. Both audible and visible alarms are activated.

c. The pump control unit shall be responsible for displaying real-time and historical operational parameters.

d. Indicator lights shall be provided for pump running state.

e. Indicator lights shall be provided for pump seal failure state.

f. Indicator lights shall be provided for pump high temperature failure state.

g. Indicator lights shall be provided for high wet well level alarm state.

A.3.2.3 Alarm Display Panel

a. Indicator lights shall be enabled when an alarm state is activated.

b. A buzzer shall sound when an alarm state is activated.

A.3.2.4 Float Switch

a. When the depth of liquid is equal to or greater than the lead pump depth, the float switch shall set a state which causes the first pump to turn on.

b. When the depth of liquid is equal to or greater than the lag pump depth, the float switch shall set a state which causes the second pump to turn on.

c. When the depth of liquid is equal to or greater than the allowable high liquid depth, the float switch shall set an alarm state.

A.3.2.5 Methane Sensor

a. When the volume of methane is equal to or greater than the high methane volume, the methane sensor shall set a state that causes the ventilation fans to turn on within 5 seconds.

b. When the volume of methane is equal to or greater than the allowable maximum methane volume, the methane sensor shall set an alarm state.

c. HighMethane represents the volume of methane that should cause the exhaust fans to turn on.

d. MaxMethane represents the volume of methane below which the wet well should be kept. Should the volume of methane be equal to or exceed MaxMethane, an alarm state is set.

e. ExhaustFan represents a Boolean quantity such that at any time *t*, the exhaust fan is either running or not running.

f. AlarmState represents a Boolean quantity such that at any time *t*, the audible and visual alarms are enabled.

$$MaxMethane > HighMethane$$

$$ExhaustFan : Time \rightarrow Bool$$

$$AlarmState : Time \rightarrow Bool$$

$$Methane \geq MaxMethane \Rightarrow ExhaustFan$$

$$Methane < MaxMethane \Rightarrow \neg ExhaustFan$$

$$Methane \geq MaxMethane \Rightarrow AlarmState$$

A.4 References

IEEE Recommended Practice for Software Requirements Specifications (IEEE Std. 830-1998).

Town of Cary North Carolina, Wet Well and Valve Vault http://www.townofcary.org/depts/dsdept/engineering/detaildrawings/ACAD-0750002-1of2.dwg, October 18, 2005.

Town of Cary North Carolina, Wet Well and Valve Vault http://www.townofcary.org/depts/dsdept/engineering/detaildrawings/ACAD-0750002-2of2.dwg, October 18, 2005.

Appendix B

Software Design for a Wastewater Pumping Station Wet Well Control System (rev. 01.01.00)

Christopher M. Garrell

B.1 Introduction

A wastewater pumping station is a component of the sanitary sewage collection system that transfers domestic sewage to a wastewater treatment facility for processing. A typical pumping station includes three components: (1) a sewage grinder, (2) a wet well, and (3) a valve vault (Figure B.1). Unprocessed sewage enters the sewage grinder unit so that solids suspended in the liquid can be reduced in size by a central cutting stack. The processed liquid then proceeds to the wet well, which serves as a reservoir for submersible pumps. These pumps then add the required energy/head to the liquid so that it can be conveyed to a wastewater treatment facility for primary and secondary treatment. The control system specification that follows describes the operation of the wet well.

B.1.1 Purpose

This specification describes the software design guidelines for the wet well control system of a wastewater pumping station. It is intended that this specification provide the basis of the software development process and is intended for use by software developers.

B.1.2 Scope

The software system described in this specification is part of a control system for the wet well of a wastewater pumping station. The control system supports an array of sensors and switches that monitor and control the operation of the wet well. The design of the wet well control system shall provide for the safety and protection of pumping station operators, maintenance personnel,

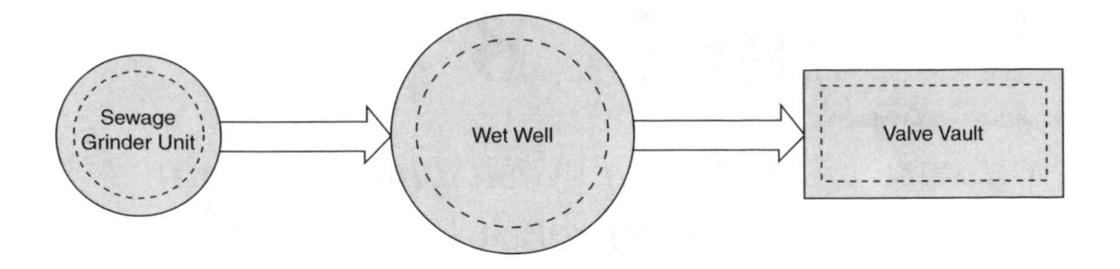

FIGURE B.1
Typical wastewater pumping station process.

and the public from hazards that may result from its operation. The control system shall be responsible for the following operations:

a. Monitoring and reporting the level of liquid in the wet well.
b. Monitoring and reporting the level of hazardous methane gas.
c. Monitoring and reporting the state of each pump and whether it is currently running or not.
d. Activating a visual and audible alarm when a hazardous condition exists.
e. Switching each submersible pump on or off in a timely fashion depending on the level of liquid within the wet well.
f. Switching ventilation fans on or off in a timely fashion depending on the concentration of hazardous gas within the wet well.

Any requirements that are incomplete are annotated with "TBD" and will be completed in a later revision of this specification.

B.1.3 Definitions, Acronyms, and Abbreviations

The following is a list of definitions for terms used in this document.

Attribute — Property of a class.

Class — A category from which object instances are created.

Message — A means of passing control from one software code unit to another software code unit because of an event.

Method — A section of software code that is associated with a class providing a mechanism for accessing the data stored in the class.

B.2 Overall Description

B.2.1 Wet Well Overview

The wet well for which this specification is intended is shown in Figure B.2. This figure has been repeated from Appendix A. A more detailed description of the wet well can be found in Appendix A.

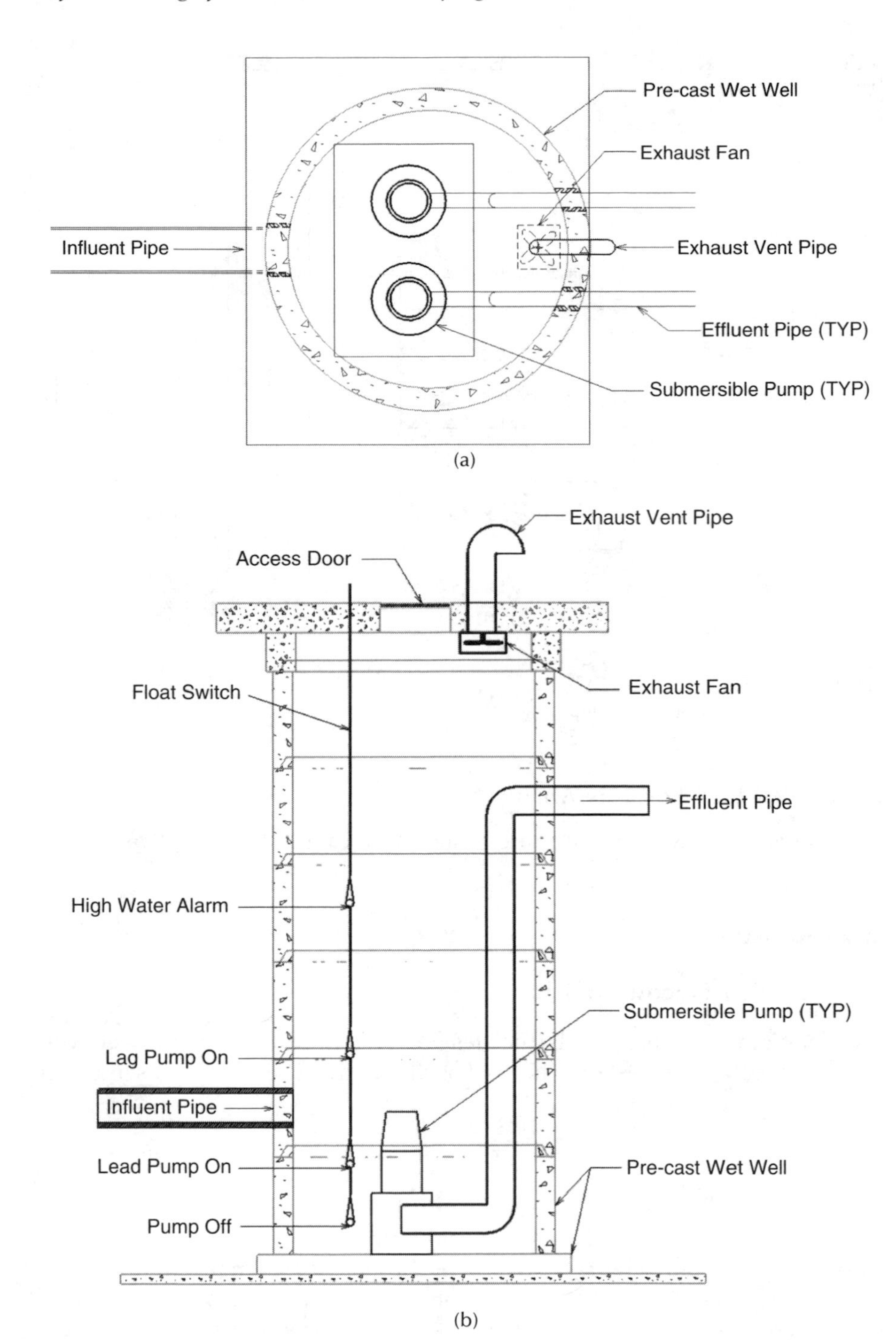

FIGURE B.2
Typical wet well. (a) Top view schematic, and (b) side sectional view.

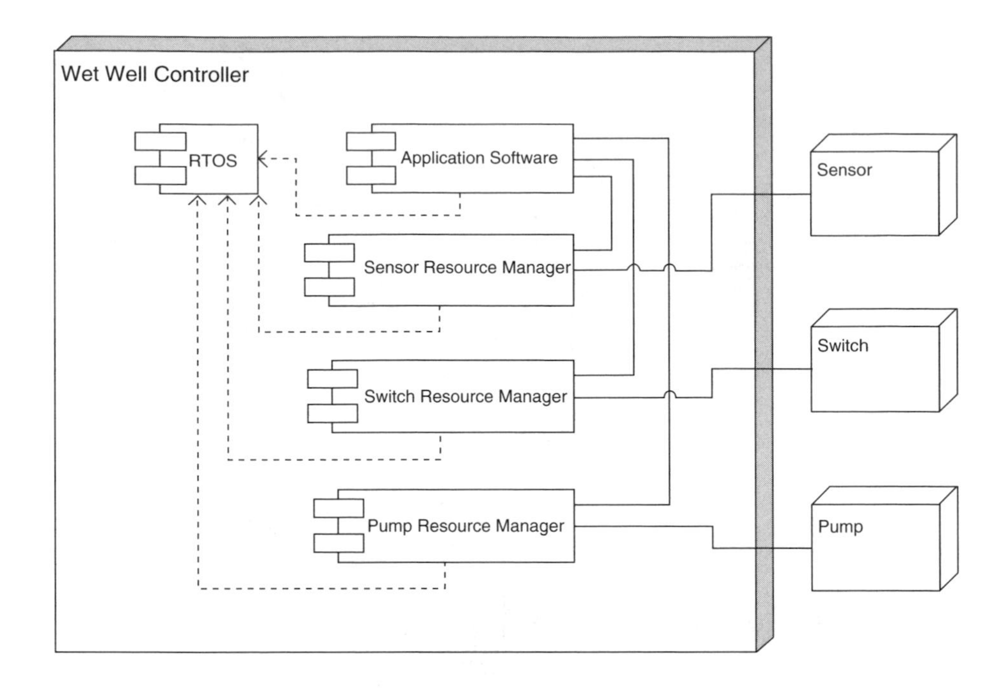

FIGURE B.3
Wet well controller software architecture.

B.2.2 Wet Well Software Architecture

The wet well software architecture is shown in Figure B.3.

B.3 Design Decomposition

The following section details the design decomposition of the wet well controller software design. This is based on the use cases presented in Appendix A.

B.3.1 Class Model

Figure B.4 describes the classes that make up the wet well control system software application. Figure B.5 describes the classes that make up the sensor state management of the wet well control system software application. Figure B.6 describes the classes that make up the process control of the wet well control system software application. Figure B.7 describes the classes that make up the resource logging control of the wet well control system software application.

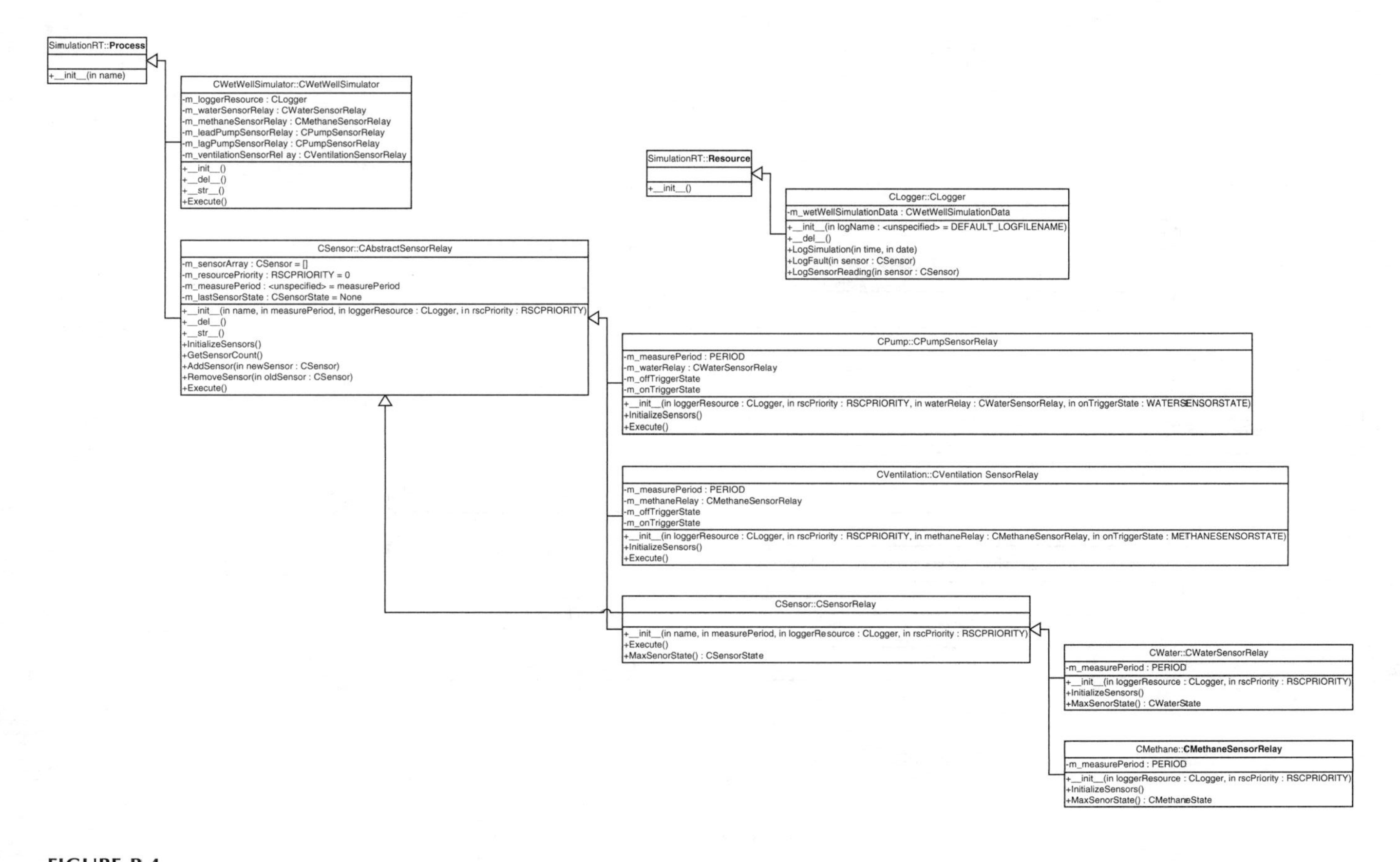

FIGURE B.4
Wet well controller class diagram.

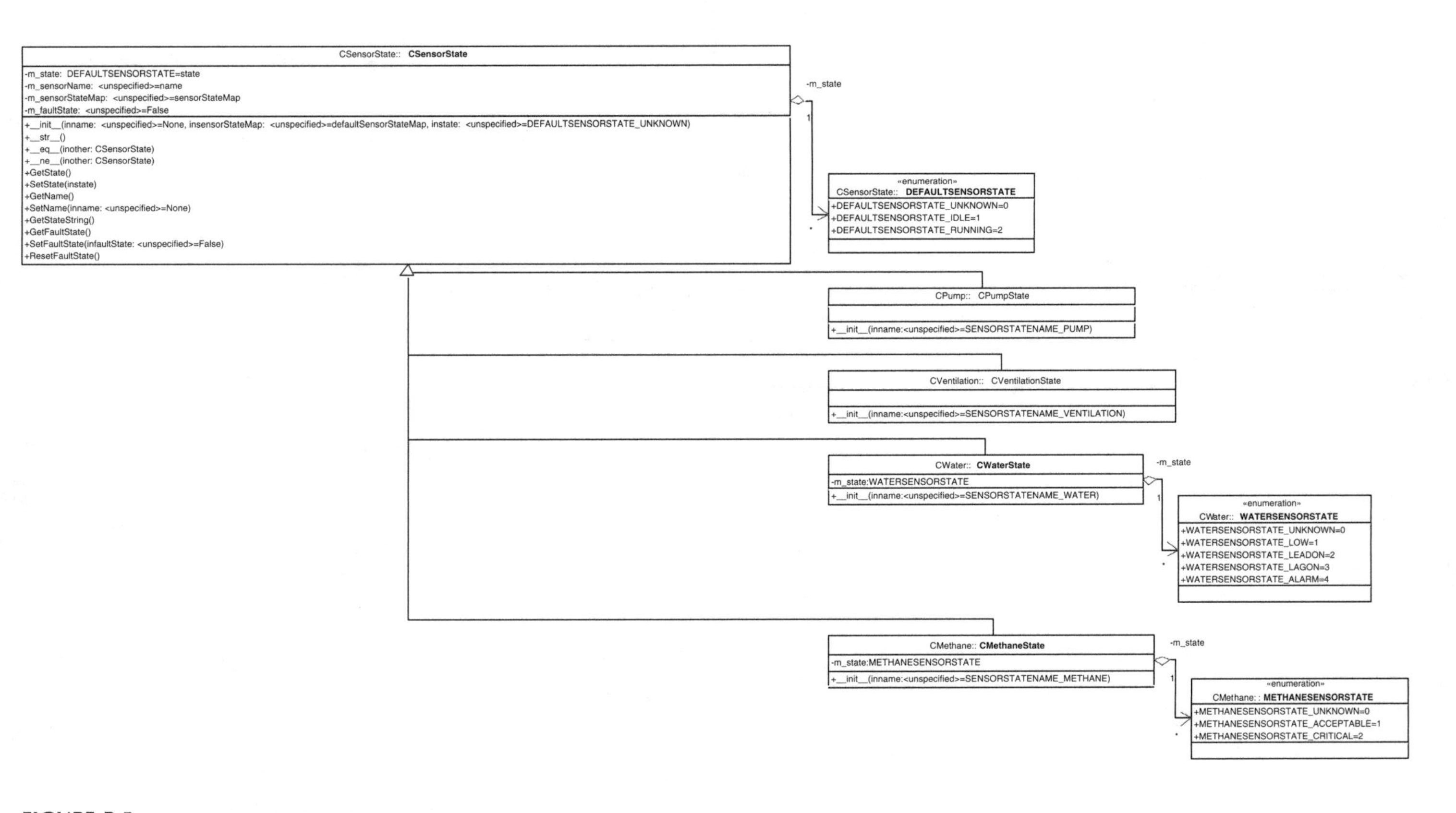

FIGURE B.5
Sensor state class diagram.

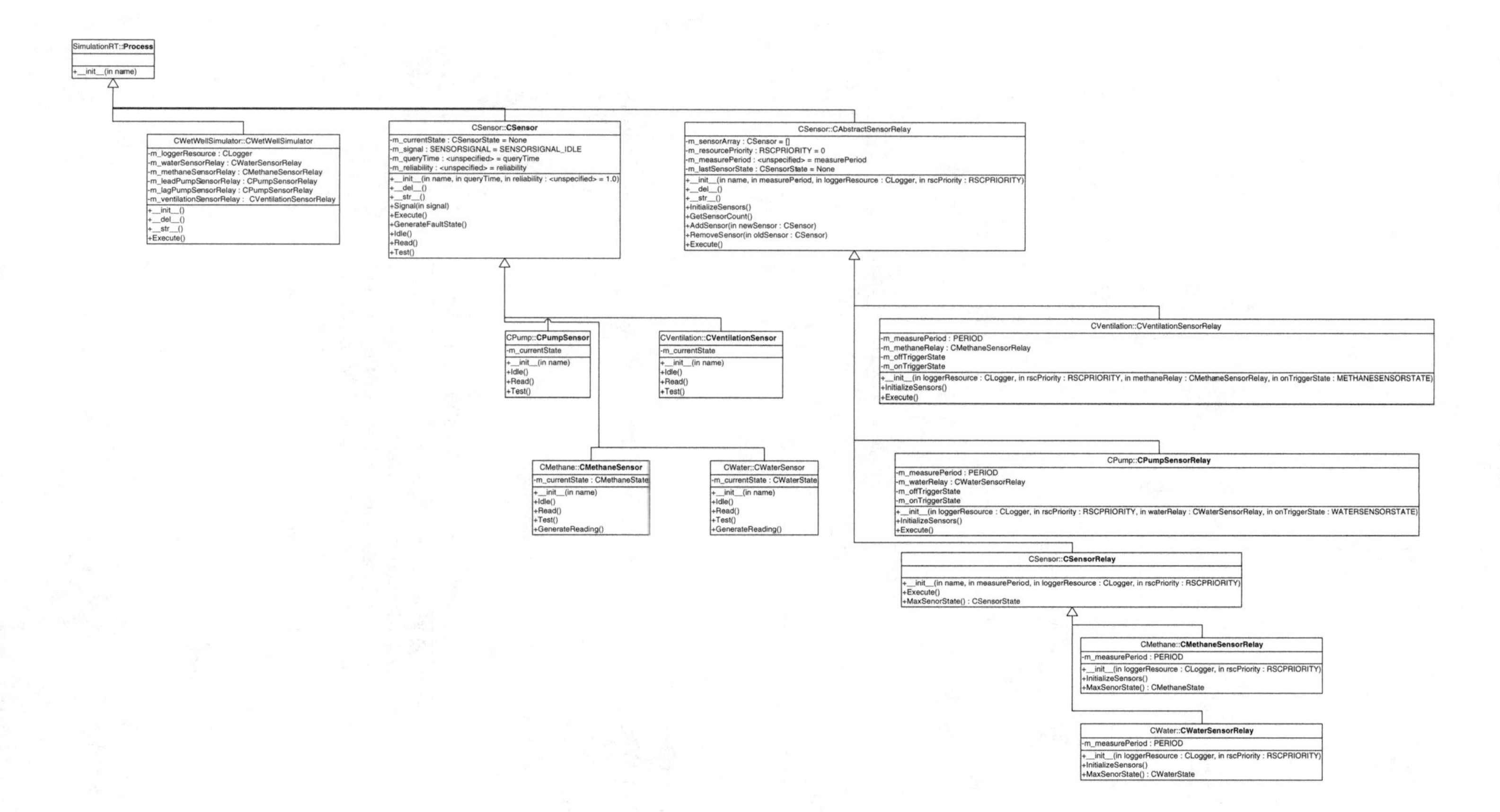

FIGURE B.6
Process control class diagram.

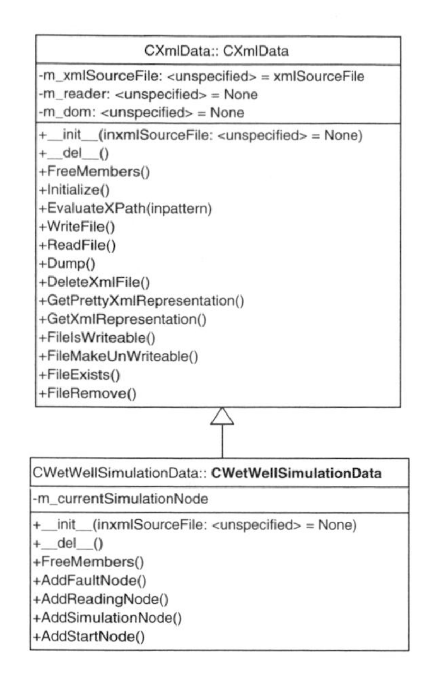

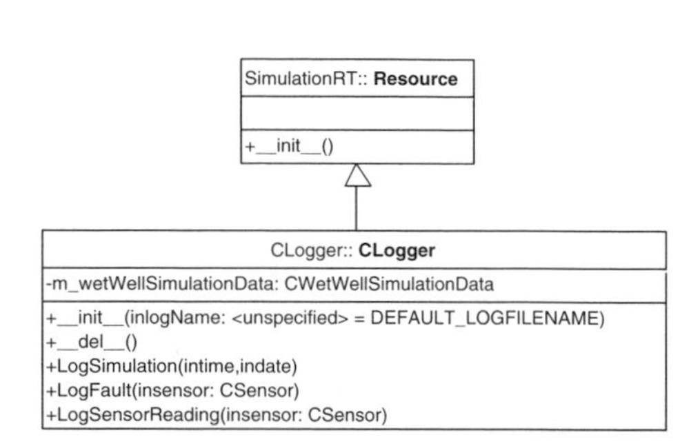

FIGURE B.7
Resource control class diagram.

B.3.2 Class Details

B.3.2.1 CWetWellSimulator

The CWetWellSimulator is responsible for the following functions (Figure B.8):

a. Initialization.
b. Instantiation of its contained objects.

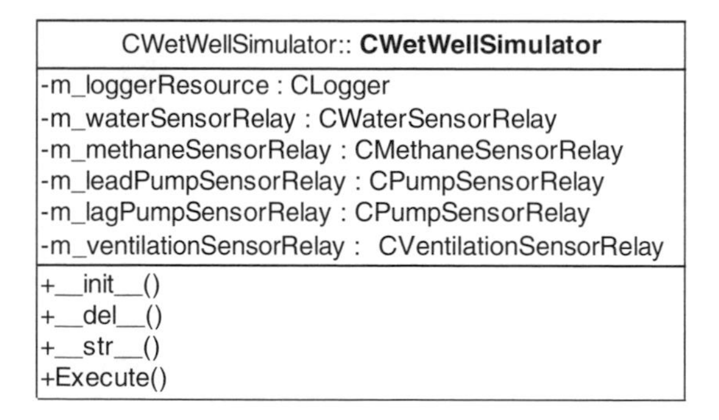

FIGURE B.8
CWetWellSimulator Class.

c. Monitoring and reporting the level of liquid in the wet well.

d. Monitoring and reporting the level of hazardous methane gas.

e. Monitoring and reporting the state of each pump and whether it is currently running.

f. Switching each submersible pump on or off in a timely fashion depending on the level of liquid within the wet well.

g. Switching ventilation fans on or off in a timely fashion depending on the concentration of hazardous gas within the wet well.

B.3.2.2 CLogger

The CLogger is responsible for the following functions (Figure B.9):

a. Initializing the simulation data logging XML file.

b. Managing the logging resource mechanism for the wet well and its sensors.

c. Logging each time the wet well control system is instantiated.

d. Logging sensor faults.

e. Logging sensor readings.

B.3.2.3 CXmlData

The CXmlData is responsible for the following functions (Figure B.10):

a. Managing generic XML content.

b. Controling XML file read and write access.

c. Adding XML elements to an XML file.

d. Adding XML attributes to an XML file.

e. Traversing XML nodes.

CLogger::**CLogger**
-m_wetWellSimulationData : CWetWellSimulationData
+__init__(in logName : <unspecified> = DEFAULT_LOGFILENAME) +__del__() +LogSimulation(in time, in date) +LogFault(in sensor : CSensor) +LogSensorReading(in sensor : CSensor)

FIGURE B.9
CLogger class.

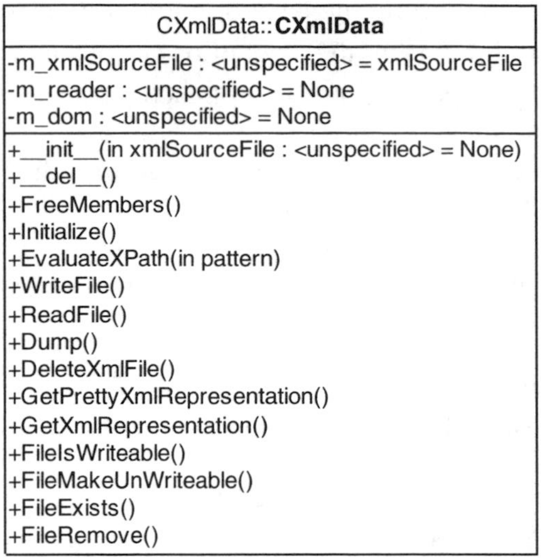

CXmlData::**CXmlData**

-m_xmlSourceFile : <unspecified> = xmlSourceFile
-m_reader : <unspecified> = None
-m_dom : <unspecified> = None

+__init__(in xmlSourceFile : <unspecified> = None)
+__del__()
+FreeMembers()
+Initialize()
+EvaluateXPath(in pattern)
+WriteFile()
+ReadFile()
+Dump()
+DeleteXmlFile()
+GetPrettyXmlRepresentation()
+GetXmlRepresentation()
+FileIsWriteable()
+FileMakeUnWriteable()
+FileExists()
+FileRemove()

FIGURE B.10
CXmlData class.

B.3.2.4 CWetWellSimulationData

The CWetWellSimulationData is responsible for the following functions (Figure B.11):

a. Representing wet well control system operation in an XML format.
b. Adding sensor fault information to XML DOM.
c. Adding sensor reading information to XML DOM.
d. Managing wet-well control system data.

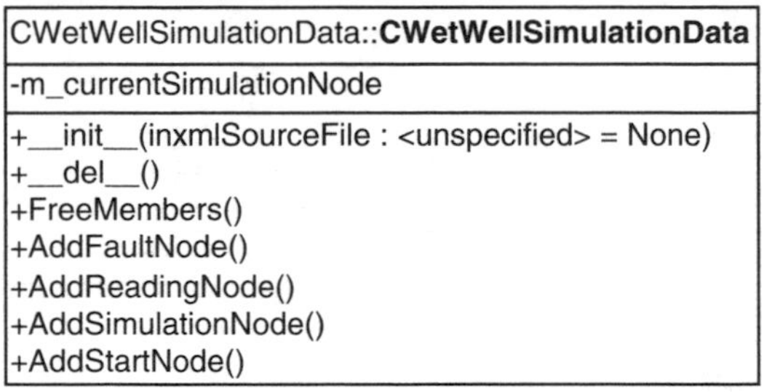

CWetWellSimulationData::**CWetWellSimulationData**

-m_currentSimulationNode

+__init__(inxmlSourceFile : <unspecified> = None)
+__del__()
+FreeMembers()
+AddFaultNode()
+AddReadingNode()
+AddSimulationNode()
+AddStartNode()

FIGURE B.11
CWetWellSimulationData class.

B.3.2.5 CSensorState

The CSensorState is responsible for the following functions (Figure B.12):

a. Maintaining the operational state of a sensor.
b. Maintaining the fault state of a sensor.

B.3.2.6 CSensor

The CSensor is responsible for the following functions (Figure B.13):

a. A process representation of a control system sensor.
b. Managing process execution.
c. Reading sensor state.
d. Storing sensor state.

B.3.2.7 CAbstractSensorRelay

The CAbstractSensorRelay is responsible for the following functions (Figure B.14):

a. Processing representations of a sensor relay control.
b. Managing process execution.
c. Managing operation of array of sensors (CSensor).
d. Providing abstract sensor array control and behavior.

B.3.2.8 CSensorRelay

The CSensorRelay is responsible for the following functions (Figure B.15):

a. Extending CAbstractSensorRelay.
b. Providing process control for sensors that take periodic reading.

B.3.2.9 CMethaneState

The CMethaneState is responsible for the following functions (Figure B.16):

a. Extending CSensorState.
b. Maintaining the operational state of a methane level sensor.
c. Maintaining the fault state of a methane level sensor.

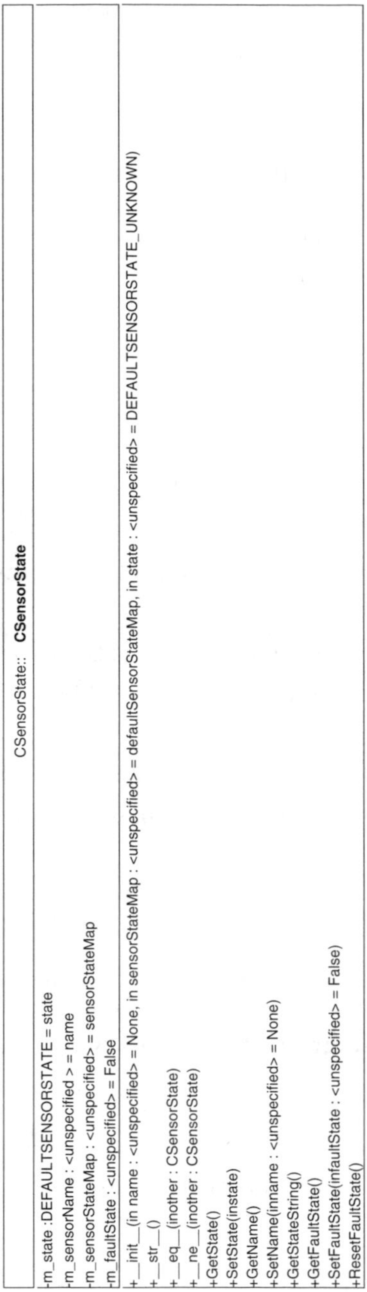

```
CSensorState:: CSensorState
-m_state :DEFAULTSENSORSTATE = state
-m_sensorName : <unspecified > = name
-m_sensorStateMap : <unspecified> = sensorStateMap
-m_faultState : <unspecified> = False
+__init__(in name : <unspecified> = None, in sensorStateMap : <unspecified> = defaultSensorStateMap, in state : <unspecified> = DEFAULTSENSORSTATE_UNKNOWN)
+__str__()
+__eq__(inother : CSensorState)
+__ne__(inother : CSensorState)
+GetState()
+SetState(instate)
+GetName()
+SetName(inname : <unspecified> = None)
+GetStateString()
+GetFaultState()
+SetFaultState(infaultState : <unspecified> = False)
+ResetFaultState()
```

FIGURE B.12
CSensorState class.

CSensor::**CSensor**
-m_currentState : CSensorState = None -m_signal : SENSORSIGNAL = SENSORSIGNAL_IDLE -m_queryTime : <unspecified> = queryTime -m_reliability : <unspecified> = reliability
+__init__(in name, in queryTime, in reliability : <unspecified> = 1.0) +__del__() +__str__() +Signal(in signal) +Execute() +GenerateFaultState() +Idle() +Read() +Test()

FIGURE B.13
CSensor class.

CSensor::**CAbstractSensorRelay**
-m_sensorArray : CSensor = [] -m_resourcePriority : RSCPRIORITY = 0 -m_measurePeriod : <unspecified> = measurePeriod -m_lastSensorState : CSensorState = None
+__init__(in name, in measurePeriod, in loggerResource : CLogger, in rscPriority : RSCPRIORITY) +__del__() +__str__() +InitializeSensors() +GetSensorCount() +AddSensor(in newSensor : CSensor) +RemoveSensor(in oldSensor : CSensor) +Execute()

FIGURE B.14
CAbstractSensorRelay class.

CSensor::**CSensorRelay**
+__init__(inname, inmeasurePeriod, inloggerResource : CLogger, inrscPriority : RSCPRIORITY) +Execute() +MaxSenorState():CSensorState

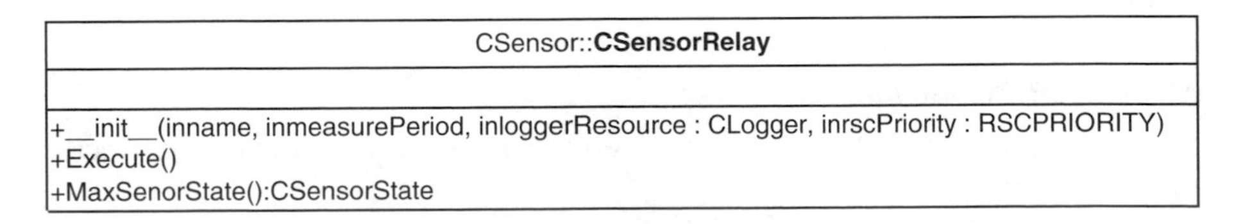

FIGURE B.15
CSensorRelay class.

CMethane::**CMethaneState**
-m_state : METHANESENSORSTATE
+__init__(in name : <unspecified> = SENSORSTATENAME_METHANE)

FIGURE B.16
CMethaneState class.

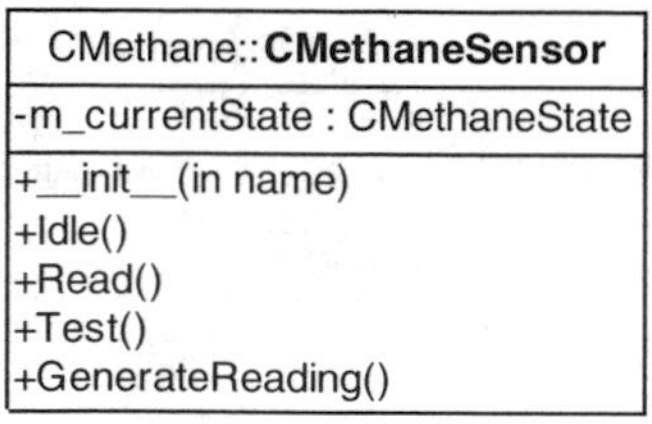

CMethane::**CMethaneSensor**
-m_currentState : CMethaneState
+__init__(in name) +Idle() +Read() +Test() +GenerateReading()

FIGURE B.17
CMethaneSensor class.

B.3.2.10 CMethaneSensor

The CMethaneSensor is responsible for the following functions (Figure B.17):

a. Extending CSensor.
b. Reading methane level.
c. Storing methane level.

Methane Elevation

Set allowable levels of methane based on a percentage (Figure B.18). Any reading above the critical level will trigger ventilation to go into an "on" state.

Methane Sensor State

This represents the sensor state based on the sensor reading (Figure B.19).

Methane Sensors Query Time

This represents the time required to make a sensor reading in seconds (Figure B.20).

Methane Sensors Reliability

This represents the reliability of the sensor in percentage of time correctly operating (Figure B.21).

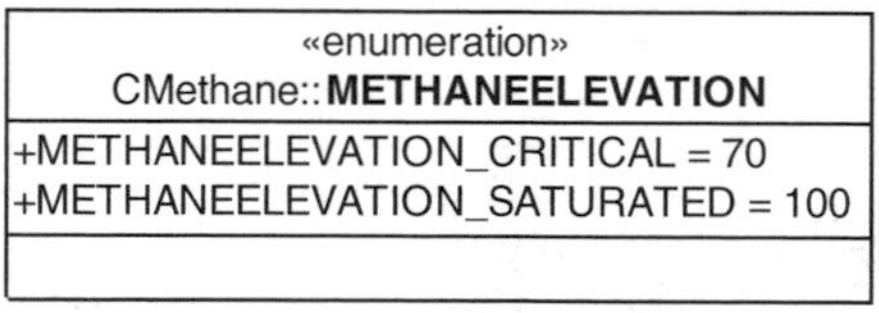

«enumeration» CMethane::**METHANEELEVATION**
+METHANEELEVATION_CRITICAL = 70 +METHANEELEVATION_SATURATED = 100

FIGURE B.18
Methane level enumerations.

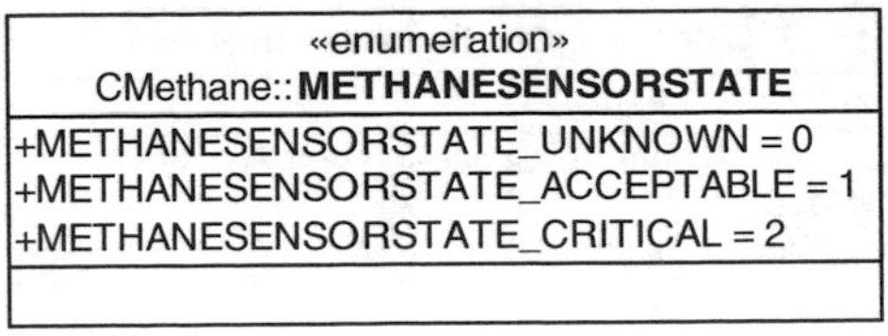

«enumeration»
CMethane:: **METHANESENSORSTATE**

+METHANESENSORSTATE_UNKNOWN = 0
+METHANESENSORSTATE_ACCEPTABLE = 1
+METHANESENSORSTATE_CRITICAL = 2

FIGURE B.19
Methane sensor state enumerations.

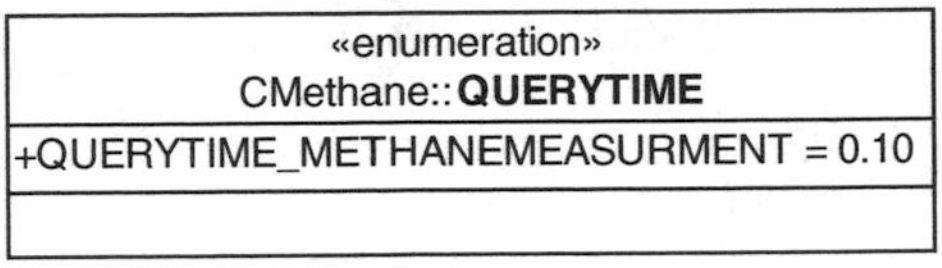

«enumeration»
CMethane:: **QUERYTIME**

+QUERYTIME_METHANEMEASURMENT = 0.10

FIGURE B.20
Methane Sensor Reading Time.

B.3.2.11 CMethaneSensorRelay

The CMethaneSensorRelay is responsible for the following functions (Figure B.22):

a. Extending CSensorRelay.
b. Providing process control for managing methane level sensors.

Number of Methane Sensors

This represents the number of sensors managed by the CMethaneSensorRelay (Figure B.23).

Methane Sensor Reading Period

This represents the time between successive sensor readings in seconds (Figure B.24).

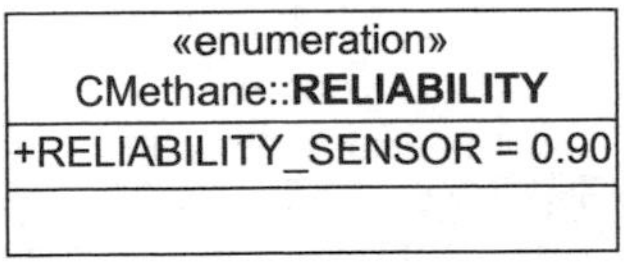

«enumeration»
CMethane::**RELIABILITY**

+RELIABILITY_SENSOR = 0.90

FIGURE B.21
Methane sensor reliability.

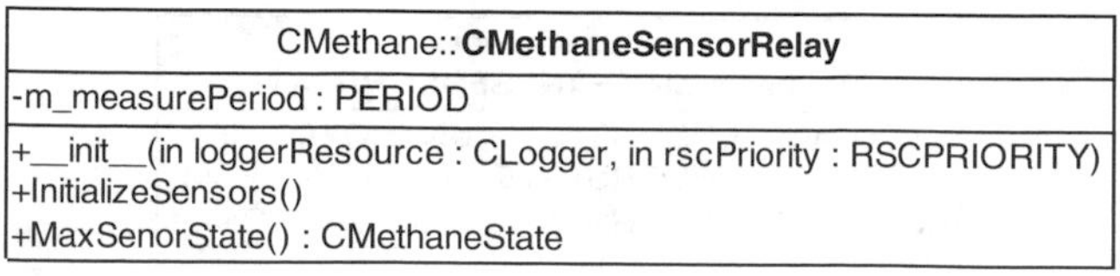

CMethane::**CMethaneSensorRelay**

-m_measurePeriod : PERIOD

+__init__(in loggerResource : CLogger, in rscPriority : RSCPRIORITY)
+InitializeSensors()
+MaxSenorState() : CMethaneState

FIGURE B.22
CMethaneSensorRelay class.

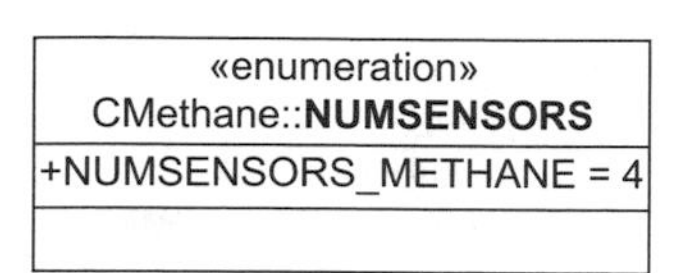

FIGURE B.23
Number of methane sensors enumeration.

B.3.2.12 CWaterState

The CWaterState is responsible for the following functions (Figure B.25):

a. Extending CSensorState.
b. Maintaining the operational state of a water level sensor.
c. Maintaining the fault state of a water level sensor.

B.3.2.13 CWaterSensor

The CWaterSensor is responsible for the following functions (Figure B.26):

a. Extending CSensor.
b. Reading water level.
c. Storing water level.

B.3.2.14 CWaterSensorRelay

The CWaterSensorRelay is responsible for the following functions (Figure B.27):

a. Extending CSensorRelay.
b. Providing process control for managing water level sensors.

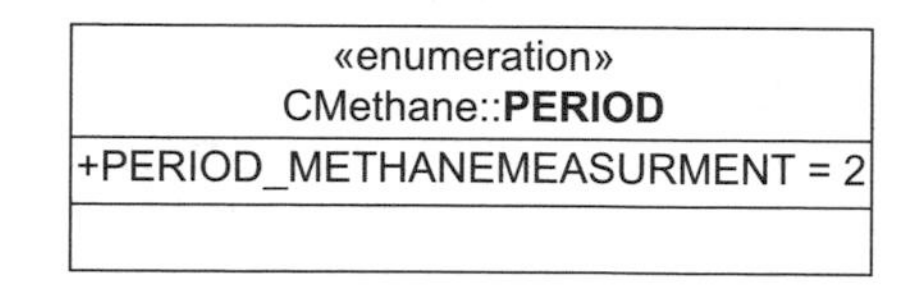

FIGURE B.24
Methane level reading period enumeration.

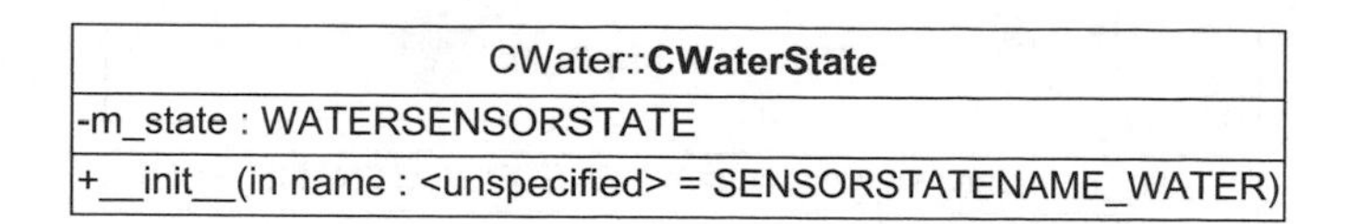

FIGURE B.25
CWaterState class.

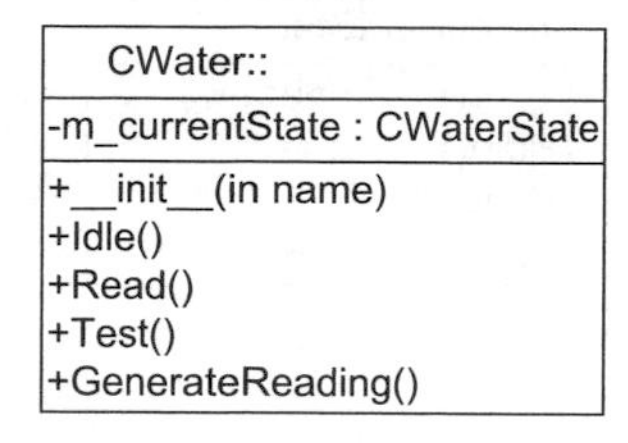

FIGURE B.26
CWaterSensor class.

B.3.2.15 CPumpState

The CPumpState is responsible for the following functions (Figure B.28):

a. Extending CAbstractSensorState.
b. Maintaining the operational state of a pump on/off sensor.
c. Maintaining the fault state of a pump on/off sensor.

B.3.2.16 CPumpSensor

The CPumpSensor is responsible for the following functions (Figure B.29):

a. Extending CSensor.
b. Reading the pump on/off state.
c. Storing the pump on/off state.

CWater::**CWaterSensorRelay**
-m_measurePeriod : PERIOD
+__init__(in loggerResource : CLogger, in rscPriority : RSCPRIORITY) +InitializeSensors() +MaxSenorState() : CWaterState

FIGURE B.27
CWaterSensorRelay class.

CPump::**CPumpState**

+__init__(in name : <unspecified> = SENSORSTATENAME_PUMP)

FIGURE B.28
CPumpState class.

CPump::**CPumpSensor**

-m_currentState

+__init__(in name)
+Idle()
+Read()
+Test()

FIGURE B.29
CPumpSensor class.

B.3.2.17 CPumpSensorRelay

The CPumpSensorRelay is responsible for the following functions (Figure B.30):

a. Extending CAbstractSensorRelay.
b. Providing process control for managing pump on/off operation depending on current water level sensor readings.

B.3.2.18 CVentilationState

The CVentilationState is responsible for the following functions (Figure B.31):

a. Extending CAbstractSensorState.
b. Maintaining the operational state of a ventilation fan sensor.
c. Maintaining the fault state of a ventilation fan sensor.

CPump:: **CPumpSensorRelay**

-m_measurePeriod : PERIOD
-m_waterRelay : CWaterSensorRelay
-m_offTriggerState
-m_onTriggerState

+__init__(in loggerResource : CLogger, in rscPriority : RSCPRIORITY, in waterRelay : CWaterSensorRelay, in onTriggerState : WATERSENSORSTATE)
+InitializeSensors()
+Execute()

FIGURE B.30
CPumpSensorRelay class.

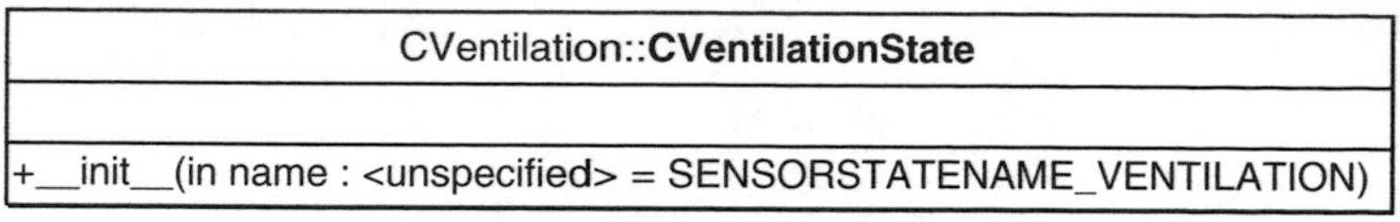

CVentilation::**CVentilationState**
+__init__(in name : <unspecified> = SENSORSTATENAME_VENTILATION)

FIGURE B.31
CVentilationState class.

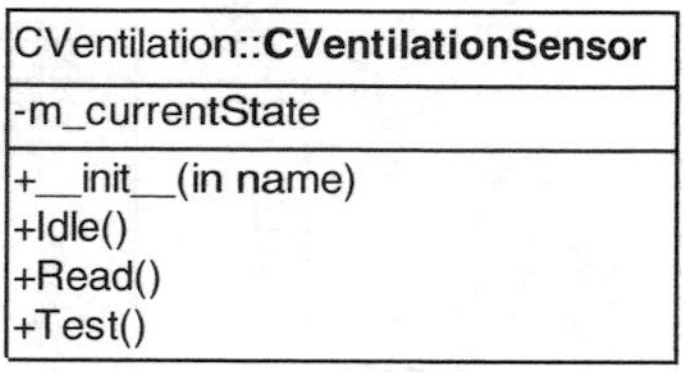

CVentilation::**CVentilationSensor**
-m_currentState
+__init__(in name) +Idle() +Read() +Test()

FIGURE B.32
CVentilationSensor class.

B.3.2.19 CVentilationSensor

The CVentilationSensor is responsible for the following functions (Figure B.32):

a. Extending CSensor.
b. Reading ventilation on/off state.
c. Storing ventilation on/off state.

B.3.2.20 CVentilationSensorRelay

The CVentilationSensorRelay is responsible for the following functions:

a. Extending CAbstractSensorRelay.
b. Providing process control for managing pump on/off operation depending on current methane level sensor readings.

B.3.3 Sequence Diagram

The sequence diagram is shown in Figure B.44.

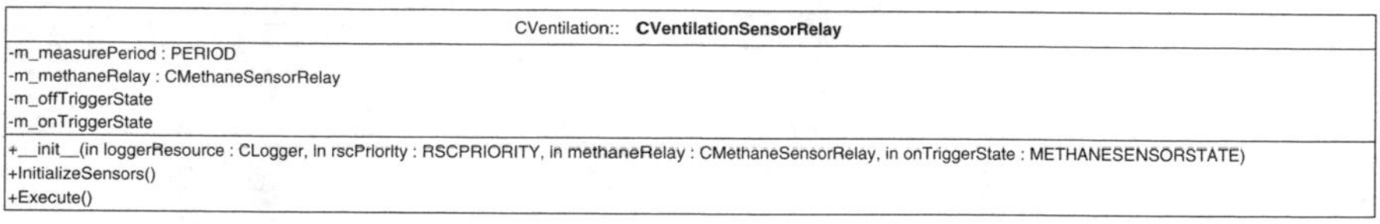

CVentilation:: **CVentilationSensorRelay**
-m_measurePeriod : PERIOD -m_methaneRelay : CMethaneSensorRelay -m_offTriggerState -m_onTriggerState
+__init__(in loggerResource : CLogger, In rscPriority : RSCPRIORITY, in methaneRelay : CMethaneSensorRelay, in onTriggerState : METHANESENSORSTATE) +InitializeSensors() +Execute()

FIGURE B.33
CVentilationSensorRelay class.

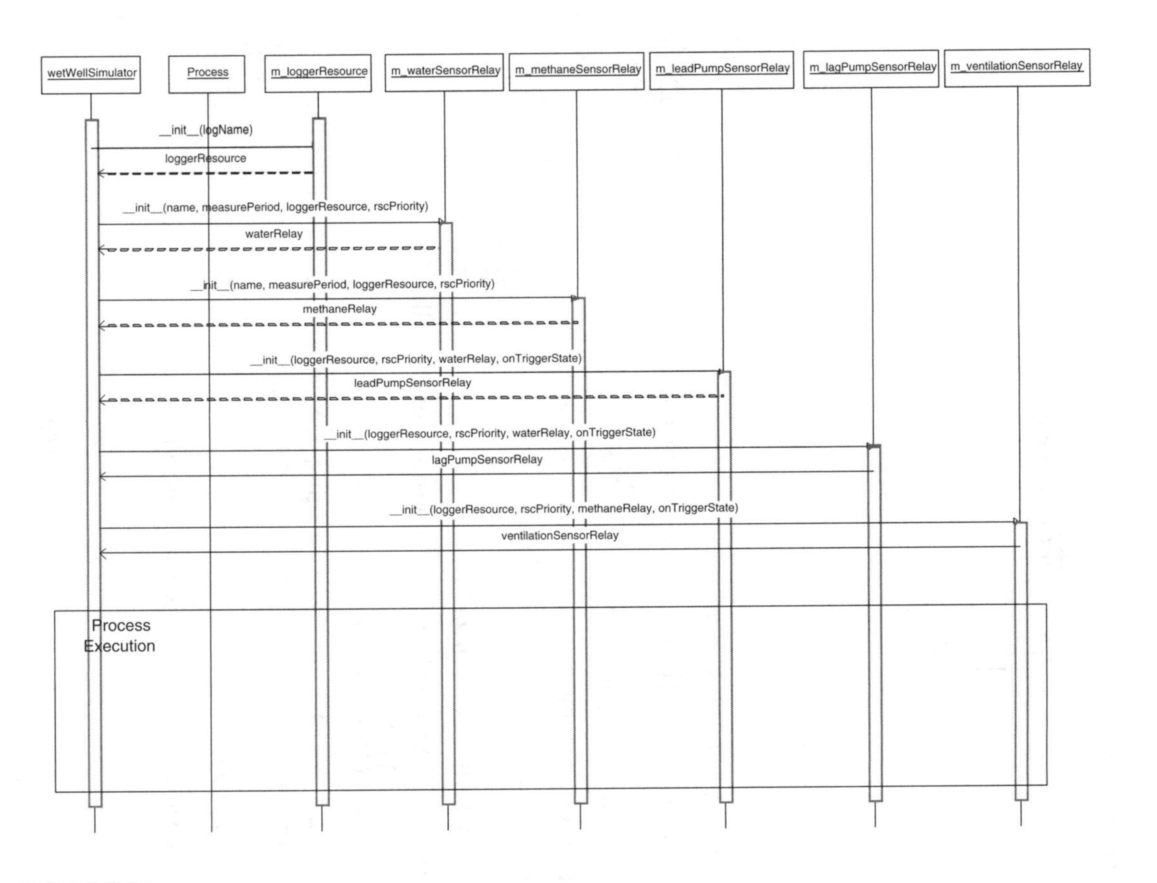

FIGURE B.34
Wet well controller sequence diagram.

B.4 References

IEEE Recommended Practice for Software Requirements Specifications (IEEE Std. 830–1998).

IEEE Recommended Practice for Software Design Descriptions (IEEE Std. 1016-1998).

Town of Cary North Carolina, Wet Well and Valve Vault http://www.townofcary.org/depts/dsdept/engineering/detaildrawings/ACAD-0750002-1of2.dwg, October 18, 2005.

Town of Cary North Carolina, Wet Well and Valve Vault http://www.townofcary.org/depts/dsdept/engineering/detaildrawings/ACAD-0750002-2 of2.dwg, October 18, 2005.

Appendix C

Object Models for a Wastewater Pumping Station Wet Well Control System

Christopher M. Garrell

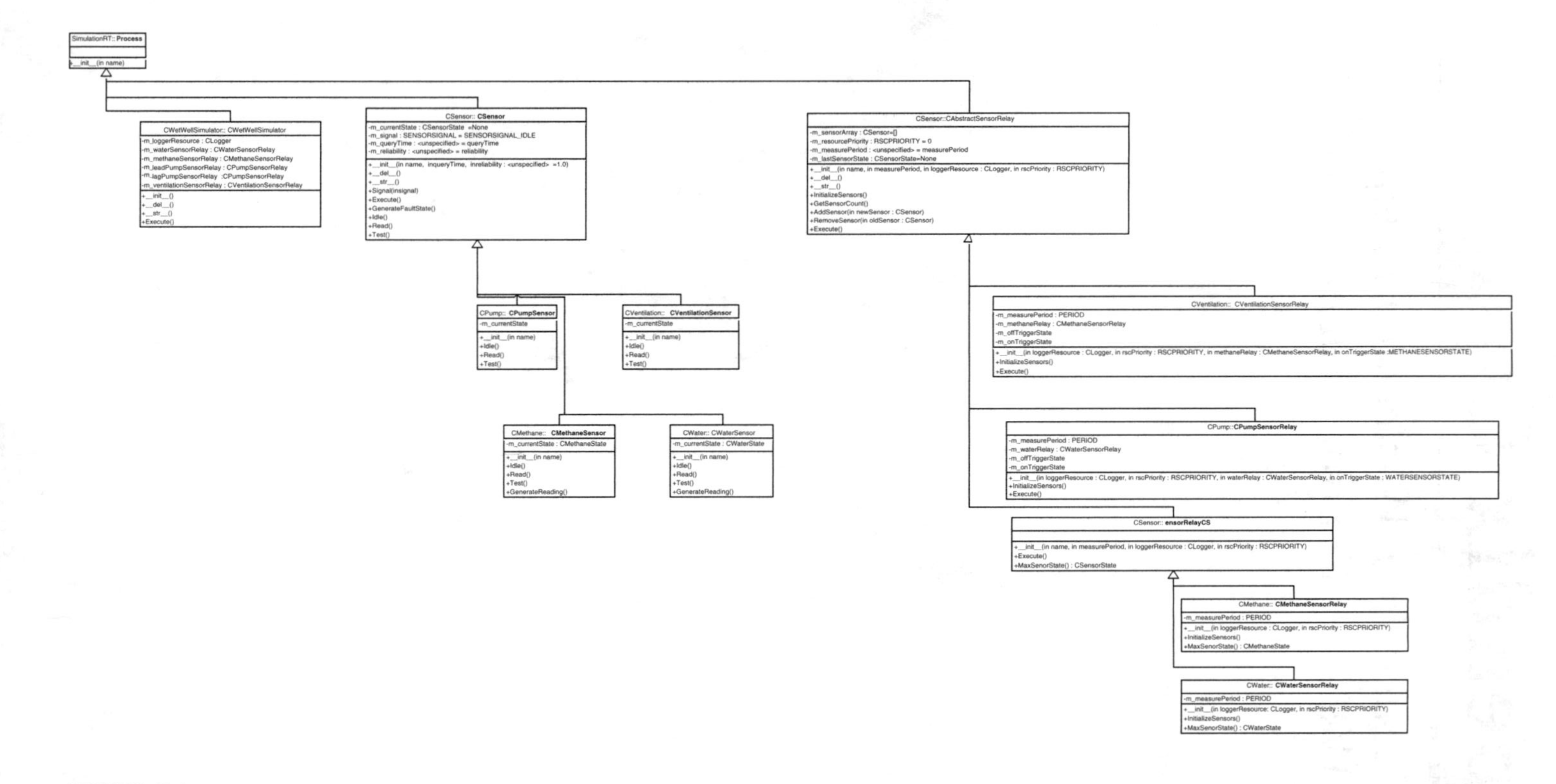

FIGURE C.1
Process control object model.

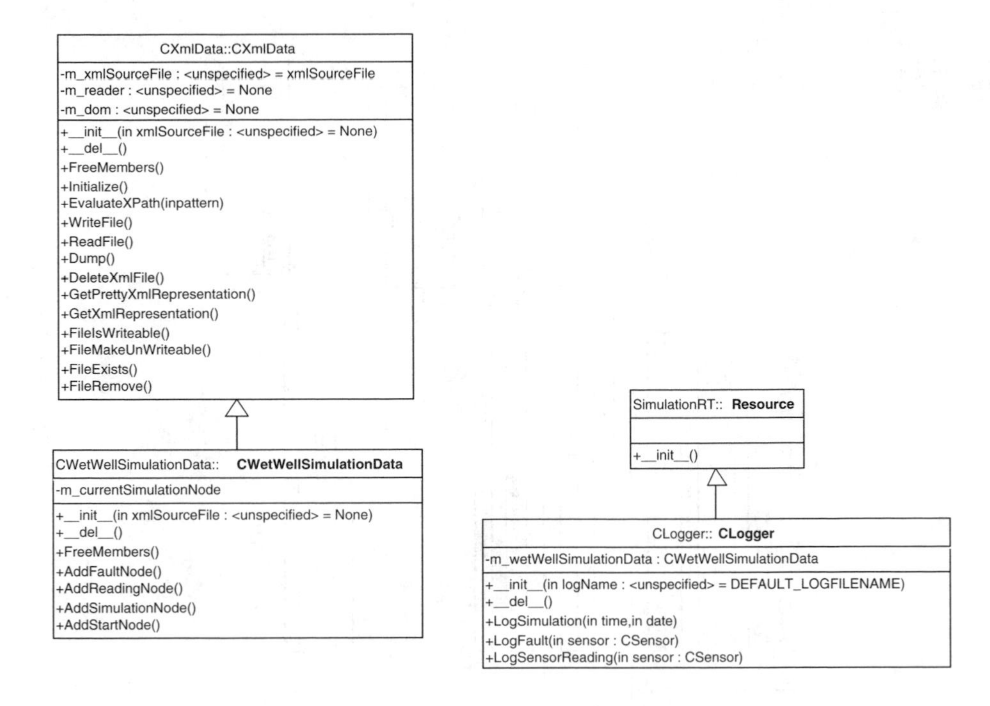

FIGURE C.2
Resource control object model.

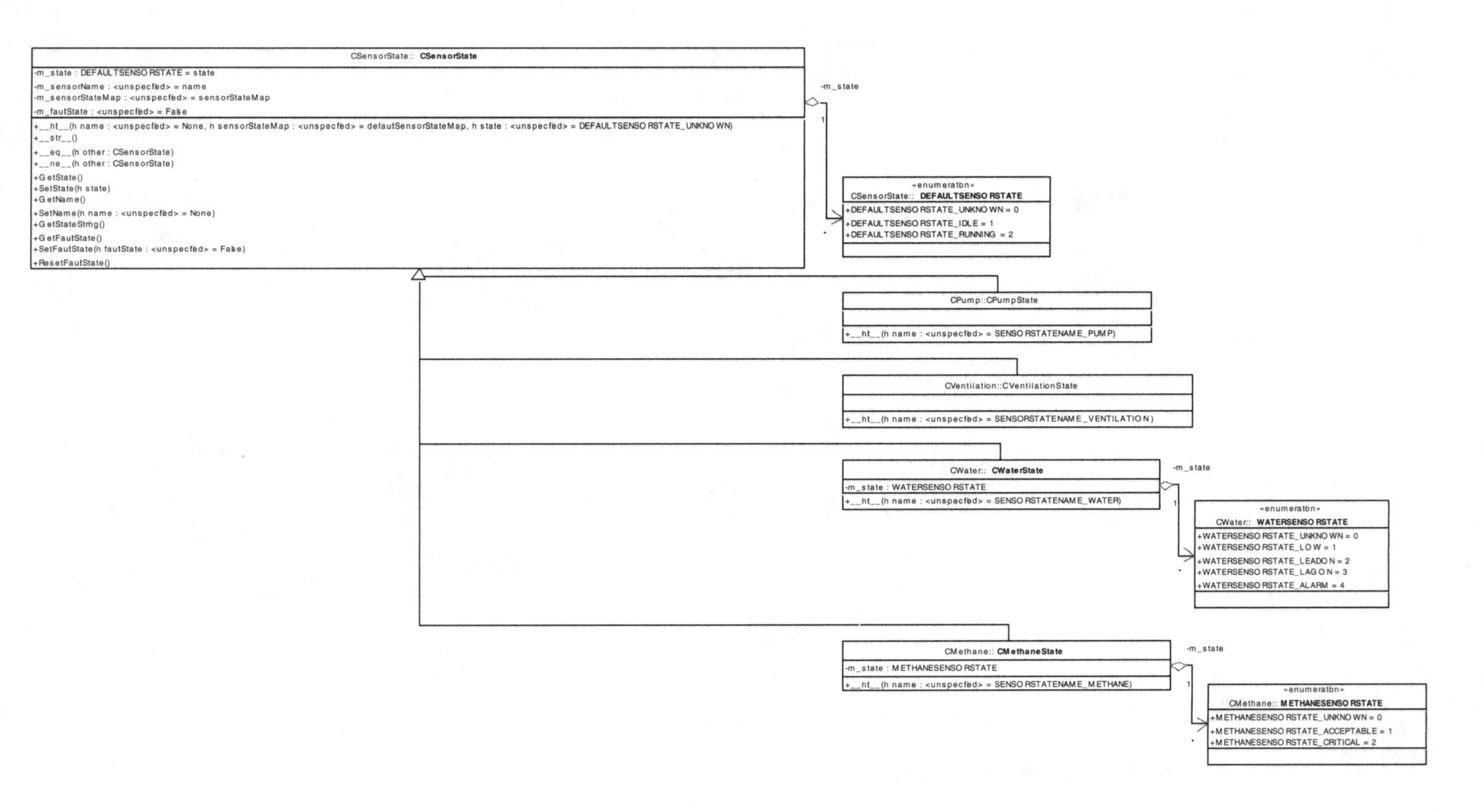

FIGURE C.3
Sensor state object model.

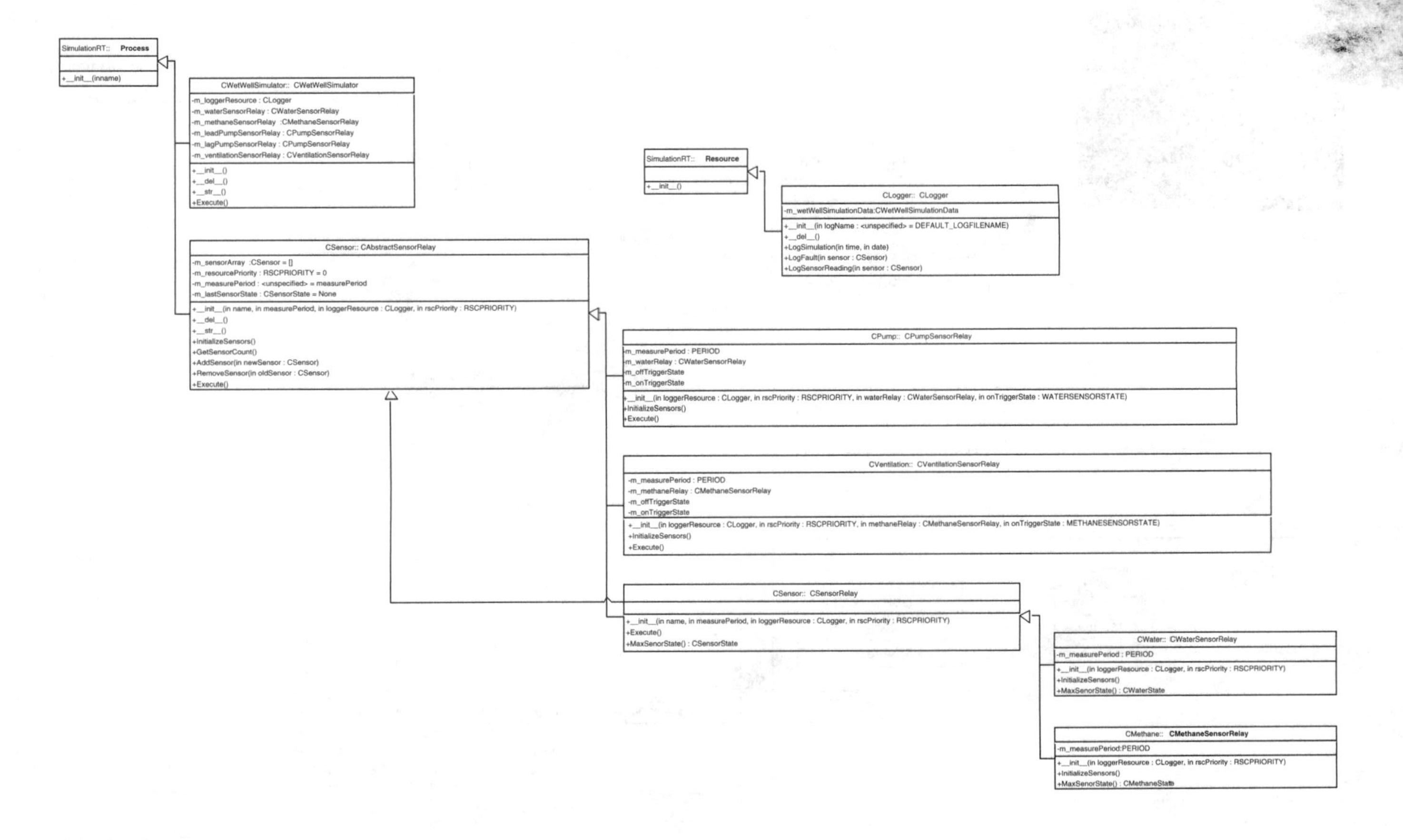

FIGURE C.4
General wet well control system object model.

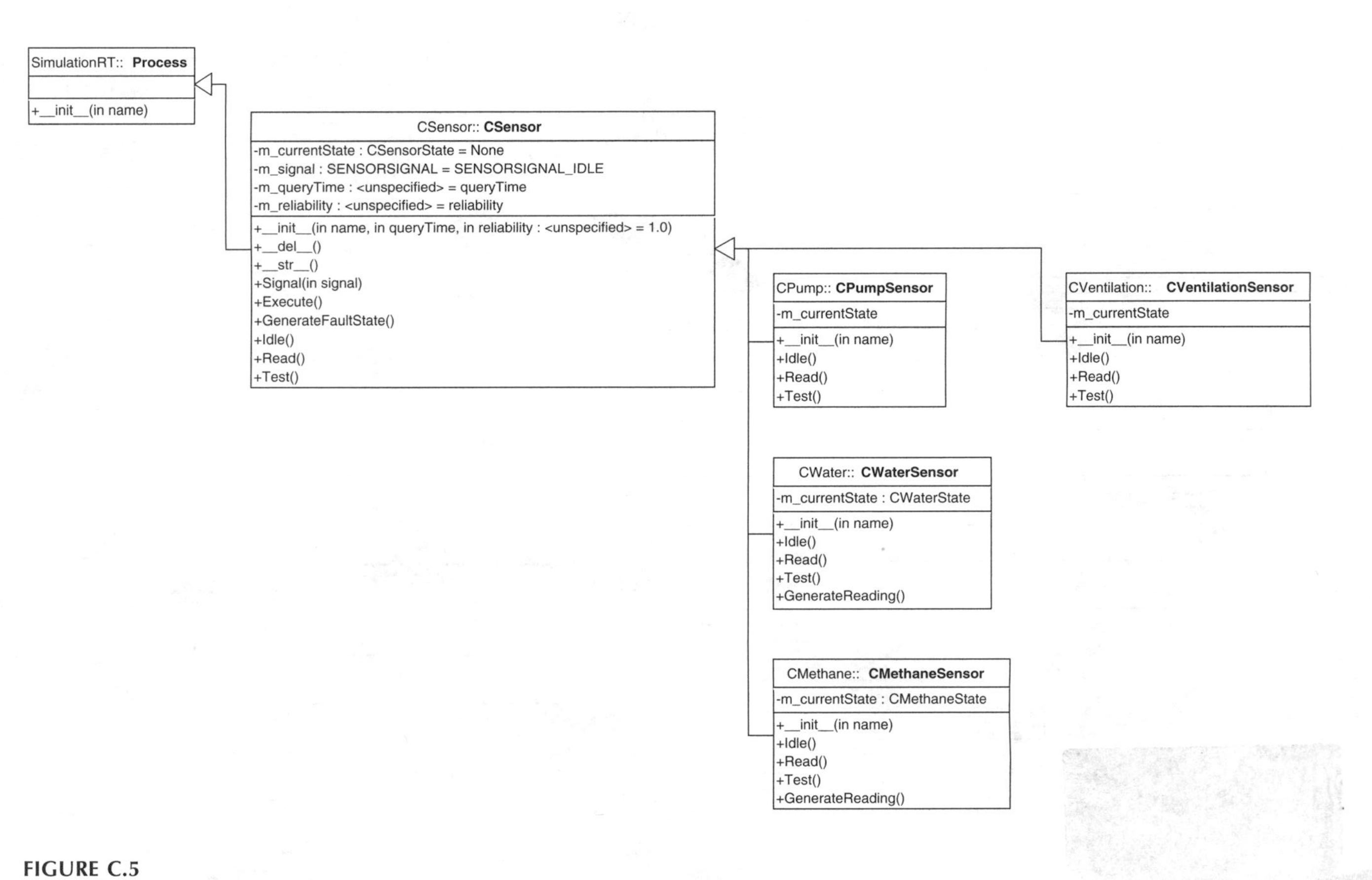

FIGURE C.5
General wet well control system object model.

Index

C

D

E

F

G

H

I

Q

R

S

U

V

W